MEI structured mathematics

Mechanics 2

J S BERRY
E GRAHAM
R PORKESS

Series Editor: Roger Porkess

MEI Structured Mathematics is supported by industry:
BNFL, Casio, GEC, Intercity, JCB, Lucas, The National Grid Company,
Texas Instruments, Thorn EMI

Hodder & Stoughton
A MEMBER OF THE HODDER HEADLINE GROUP

Acknowledgements

We are grateful to the following companies, institutions and individuals who have given permission to reproduce photographs in this book. Every effort has been made to trace and acknowledge ownership of copyright. The publishers will be glad to make suitable arrangements with any copyright holders whom it has not been possible to contact.

J Allan Cash Photolibrary pages 2, 51, 126 (right),
Colin Taylor Productions pages 13; 16 (all 3), 20 (both), 48 (both), 90, 135
Coloursport pages 96, 102,
Imperial College page 134
Manchester Olympic Games Executive Committee Ltd page 126 (left)
Northern Counties Buses page 42
Unilab page 129
Page 71© 1961 M.C. Escher Foundation-Baarn-Holland. All rights reserved.

Cataloguing in Publication Data is available from the British Library

ISBN 0 340 573015

First published 1994
Impression number 10 9 8 7 6 5 4 3 2 1
Year 1998 1997 1996 1995 1994

Typeset by Multiplex Techniques Ltd.
Printed in Great Britain for Hodder & Stoughton Educational, a division of Hodder Headline Plc, 338 Euston Road, London NW1 3BH by Thomson Litho Ltd.

MEI Structured Mathematics

Mathematics is not only a beautiful and exciting subject in its own right but also one that underpins many other branches of learning. It is consequently fundamental to the success of a modern economy.

MEI Structured Mathematics is designed to increase substantially the number of people taking the subject post-GCSE, by making it accessible, interesting and relevant to a wide range of students.

It is a credit accumulation scheme based on 45 hour components which may be taken individually or aggregated to give:

3 components	AS Mathematics
6 components	A Level Mathematics
9 components	A Level Mathematics + AS Further Mathematics
12 components	A Level Mathematics + A Level Further Mathematics

Components may alternatively be combined to give other A or AS certifications (in Statistics, for example) or they may be used to obtain credit towards other types of qualification.

The course is examined by the Oxford and Cambridge Schools Examination Board, with examinations held in January and June each year.

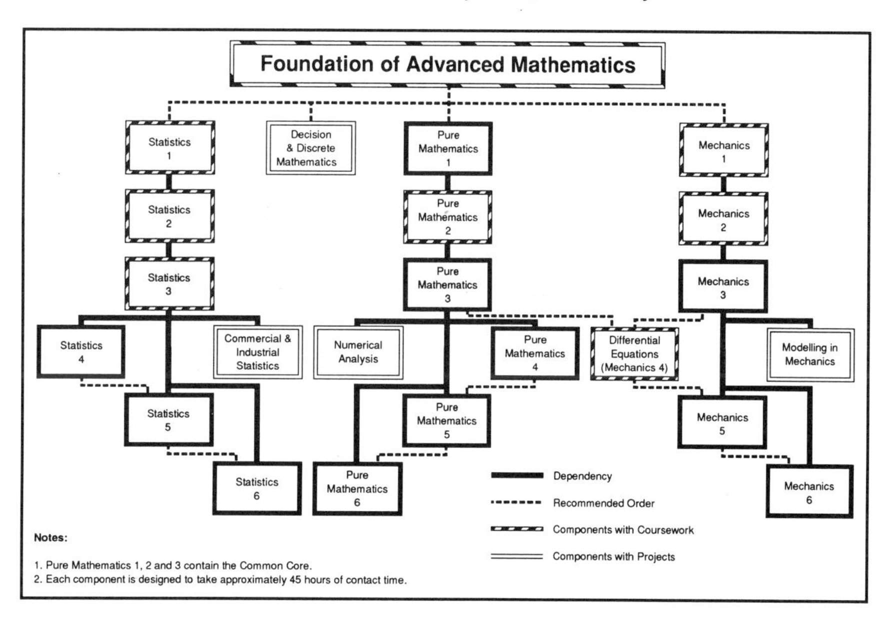

This is one of the series of books written to support the course. Its position within the whole scheme can be seen in the diagram above.

Mathematics in Education and Industry is a curriculum development body which aims to promote the links between Education and Industry in Mathematics at secondary school level, and to produce relevant examination and teaching syllabuses and support material. Since its foundation in the 1960s, MEI has provided syllabuses for GCSE (or O Level), Additional Mathematics and A Level.

For more information about MEI Structured Mathematics or other syllabuses and materials, write to MEI Office, Monkton Combe, Bath BA2 7HG.

Introduction

This is the second in a series of books written to support the Mechanics Components of MEI Structured Mathematics, but you may also use them for an independent course of study in the subject. Throughout the series emphasis is placed on understanding the basic principles of Mechanics and the process of modelling the real world, rather than on mere routine calculations.

This book builds on the concepts introduced in Mechanics 1, giving a deeper treatment of some of the ideas you met there, like friction and centre of mass. Previous work on forces is applied to pin-jointed structures. The book also introduces a number of new concepts: the rigid body model, energy and momentum.

Some examples of everyday applications are covered in the worked examples, many more in the exercises. Working through these exercises is an important part of learning the subject. Not only will it help you to appreciate the wide variety of situations to which the principles of Mechanics can be applied. It will also build your confidence in selecting and using the principles relevant to any situation.

Mechanics is about modelling the real world, and this involves observing what is going on around you and asking questions about it. This book contains several experiments and investigations for you to carry out; these are designed to enable you to bridge the gap between pen and paper exercises and the world you see about you.

We would like to thank Sharon Ward who typed the manuscript and all those who read through earlier drafts of this book, providing us with many helpful suggestions.

John Berry, Ted Graham, Roger Porkess.

Contents

1 A model for friction

Theories do not have to be "right" to be useful.

Alvin Toffler

This statement about a road accident was offered to a magistrates court by a solicitor.

'Briefly the circumstances of the accident are that our client was driving his Porsche motor car. He had just left work at the end of the day. He was stationary at the junction with Plymouth Road when a motorcyclist travelling down from Tavistock lost control of his motorcycle due to excessive speed, and collided with the front offside of our client's motor car.

'The motorcyclist was braking when he lost control and left a 26 metre skid mark on the road. Our advice from an expert witness is that the motorcyclist was exceeding the speed limit of 30 mph'.

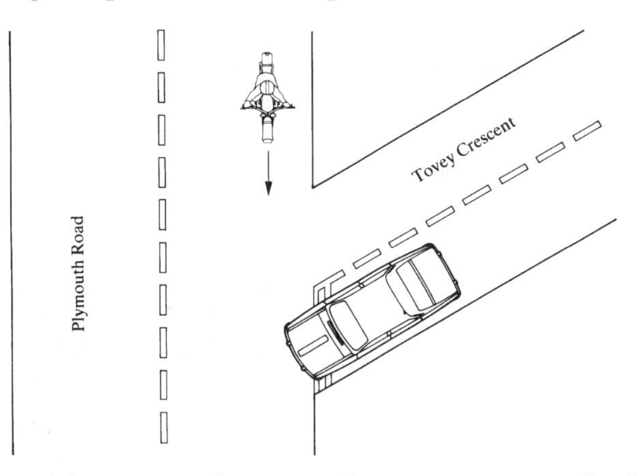

It is the duty of the court to determine the innocence or guilt of the motorcyclist. The key question is *Was the motorcyclist exceeding the speed limit?*

The car driver's solicitor is clearly implying that the motorcyclist was to blame. Can the applied mathematician help to sort out the problem? Is it possible to deduce the speed of the motorcyclist from the skid mark?

For Discussion

Discuss this situation and make a list of the important factors that you would need to consider in modelling it.

A model for friction

Clearly the key information in this situation is provided by the skid marks. To interpret it, you need a model for how friction works; in this case the friction between the motorcycle's tyres and the road.

The laws of friction are usually attributed to Coulomb. As a result of experimental work, he formulated the following model for friction between two surfaces.

1. Friction always opposes relative motion between two surfaces in contact.
2. Friction is independent of the relative speed of the surfaces.
3. The magnitude of the frictional force has a maximum which depends on the normal reaction between the surfaces and on the roughness of the surfaces in contact.
4. If there is no sliding between the surfaces

$$F \leq \mu R \quad \text{where } F = \text{force due to friction}$$
$$R = \text{normal reaction}$$
$$\mu = \text{coefficient of friction}$$

If sliding occurs

$$F = \mu R$$

According to Coulomb's model, μ is a constant for any pair of surfaces.

Typical values and ranges of values for the coefficient of friction μ are given in this table.

Surfaces in contact	μ
wood sliding on wood	0.2 – 0.6
metal sliding on metal	0.15 – 0.3
normal tyres on dry road	0.8
racing tyres on dry road	1.0
sandpaper on sandpaper	2.0
skis on snow	0.02

How fast was the motorcyclist going?

You can now proceed with the problem. As an initial model, you might make the following assumptions:

1. that the road is level
2. that the motorcycle was at rest just as it hit the car. (Obviously it was not, but this assumption allows you to estimate a minimum initial speed for the motorcycle);
3. that the motorcycle and rider may be treated as a particle, subject to Coulomb's laws of friction with $\mu = 0.8$ (i.e. dry road conditions).

The calculation then proceeds as follows.

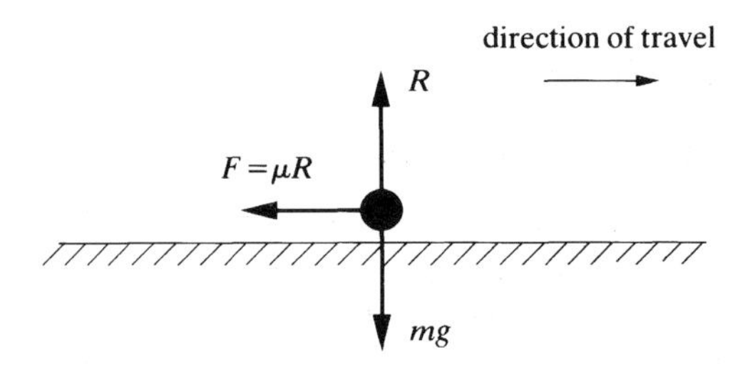

Figure 1.1

Taking the direction of travel as positive, let the motorcycle and rider have acceleration a and mass m. You have probably realised that the acceleration will turn out to be negative. The forces are shown in figure 1.1.

Applying Newton's second law:

perpendicular to the road, since there is no vertical acceleration we have

$$R - mg = 0;$$

parallel to the road, there is a constant force $-\mu R$ from friction, so we have

$$-\mu R = ma.$$

Solving for a gives

$$a = -\frac{\mu R}{m} = -\frac{\mu mg}{m} = -\mu g.$$

Taking $g = 10$ ms^{-2} and $\mu = 0.8$ gives $a = 8$ ms^{-2}.

The constant acceleration equation

$$v^2 = u^2 + 2as$$

can be used to to calculate the initial speed of the motorcycle. Substituting $s = 26$, $v = 0$ and $a = 8$ gives

$$u = \sqrt{2 \times 8 \times 26} = 20.4 \text{ in ms}^{-1}$$

This figure can be converted to miles per hour (using the fact that 1 mile = 1600 m):

$$\text{speed} = \frac{20.4 \times 3600}{1600} \text{ mph}$$

$$= 45.6 \text{ mph.}$$

So this first simple model suggests that the motorcycle was travelling at a speed of at least 45.6 mph before skidding began.

For Discussion

Is this a good model?

Discuss how good this model is and whether you would be confident in offering the answer as evidence in court. Look carefully at the three assumptions. What effect do they have on the estimate of the initial speed?

Modelling with friction

Whilst there is always some frictional force between two sliding surfaces its magnitude is often very small. In such cases we ignore the frictional force and describe the surfaces as *smooth*.

In situations where frictional forces cannot be ignored we describe the surface(s) as *rough*. Coulomb's Law is the standard model for dealing with such cases.

Frictional forces are essential in many ways. For example, a ladder leaning against a wall would always slide if there were no friction between the foot of the ladder and the ground. The absence of friction in icy conditions causes difficulties for road users: pedestrians slip over, cars and motorcycles skid.

Remember that friction never causes motion; it always opposes sliding motion.

HISTORICAL NOTE

Charles Augustin de Coulomb was born in Angoulême in France in 1736 and is best remembered for his work on electricity rather than for that on friction. Coulomb's Law concerns the forces on charged particles and was determined using a torsion balance. The unit for electric charge is named after him.

Coulomb was a military engineer and worked for many years in the West Indies, eventually returning to France in poor health not long before the revolution. He worked in many fields, including the elasticity of metal, silk fibres and the design of windmills. He died in Paris in 1806.

EXAMPLE

A rope is attached to a crate of mass 70 kg at rest on a flat surface. The coefficient of friction between the floor and the crate is 0.6. Find the maximum force that the rope can exert on the crate without moving it.

Solution:

The forces acting on the crate are shown in the diagram.

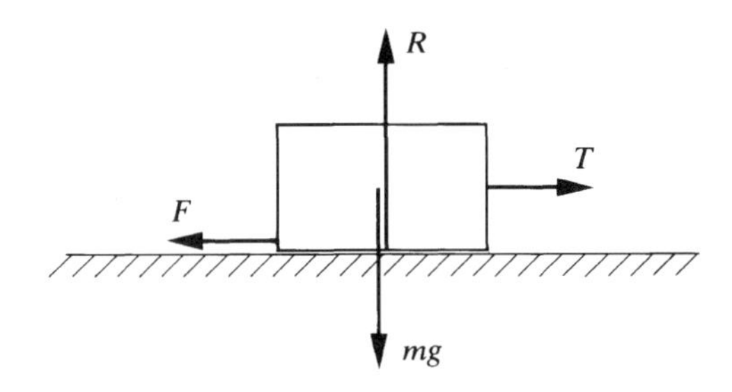

Since the crate does not move, it is in equilibrium.

Horizontal forces: $T = F$

Vertical forces: $R = mg$

$= 70g = 686$ (N)

The law of friction states that:

$F \leq \mu R$ for objects at rest.

So in this case $F \leq 0.6 \times 686\,g$

The maximum value of F is therefore 412 (N). As the tension in the rope and the force of friction are the only forces which have horizontal components, the crate will remain in equilibrium unless the tension in the rope is greater than 412 N.

EXAMPLE

The diagram shows a block of mass 5 kg resting on a rough table and connected by light inextensible strings passing over smooth pulleys to masses of 2 kg and 7 kg. The coefficient of friction between the block and the table is 0.4.

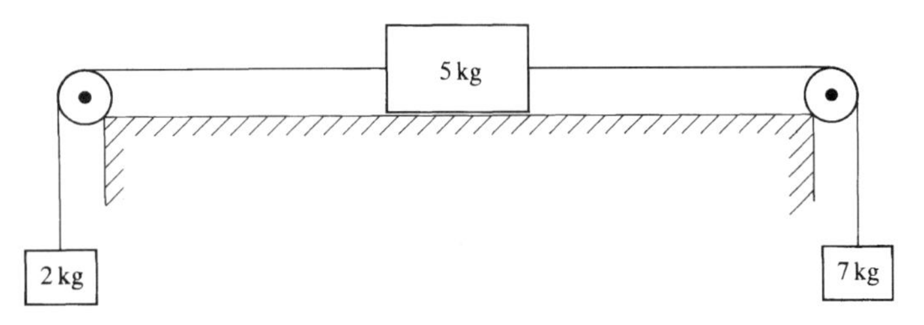

(i) Draw a diagram showing the forces acting on the three blocks and the direction of the system's acceleration.

(ii) Find the acceleration of the system and the tensions in the strings.

Solution:

(i)

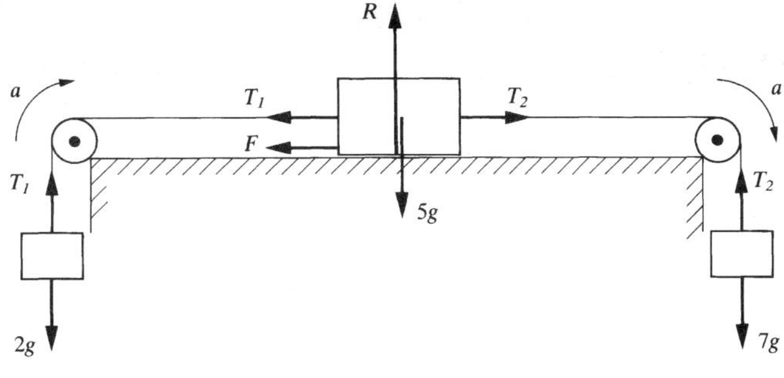

(ii) Since sliding occurs, $F = \mu R$

For the 5 kg block,

Vertically: $R = 5g$ (since the block has no vertical acceleration)

Horizontally: $T_2 - T_1 - \mu R = 5a$

Substituting for μ and R, the second equation gives:

$$T_2 - T_1 - 2g = 5a \quad (1)$$

For the 2 kg block, vertically:

$$T_1 - 2g = 2a \quad (2)$$

For the 7 kg block, vertically:

$$7g - T_2 = 7a$$

Adding (1), (2) and (3), $\quad 3g = 14a \quad (1)$

$$a = \frac{3}{14}g = 2.1$$

Back-substituting gives $T_1 = 23.8, T_2 = 53.9$

The acceleration is 2.1 ms^{-2} the tensions are 23.8 N and 53.9 N.

EXAMPLE

Angus is pulling a sledge of mass 12 kg at steady speed across level snow by means of a rope which makes an angle of 20° with the horizontal. The coefficient of friction between the sledge and the ground is 0.15. What is the tension in the rope?

Solution

Since the sledge is travelling at steady speed, the forces acting on it are in equilibrium. They are shown in the diagram.

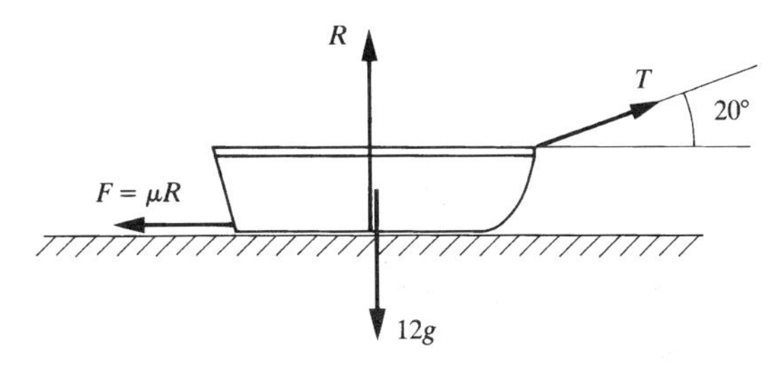

Horizontally: $T \cos 20° = \mu R$

$= 0.15\,R$

Vertically: $T \sin 20° + R = 12\,g$

$R = 12 \times 9.8 - T \sin 20°$

Combining these gives

$$T \cos 20° = 0.15\,(12 \times 9.8 - T \sin 20°)$$

$$T\,(\cos 20° + 0.15 \sin 20°) = 0.15 \times 12 \times 9.8$$

$$T = 17.8$$

The tension is 17.8 N.

EXAMPLE

A ski slope is designed for beginners. Its angle to the horizontal is such that skiers will either remain at rest on the point of moving or, if they are pushed off, move at constant speed. The coefficient of friction between the skis and the slope is 0.35. Find the angle that the slope makes with the horizontal.

Solution

The diagram shows the forces on the skier.

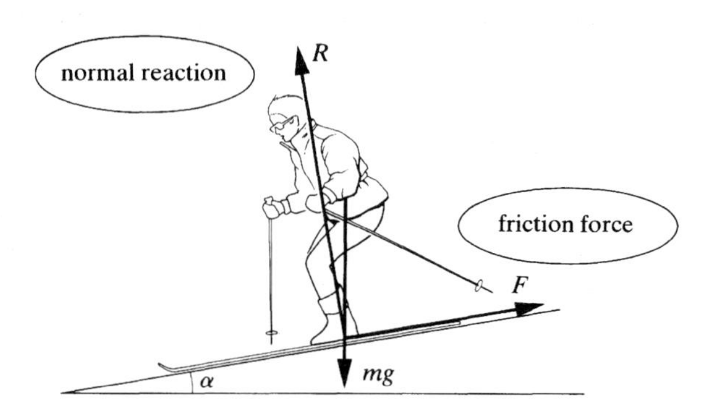

The weight mg can be resolved into components $mg \cos \alpha$ perpendicular to the slope and $mg \sin \alpha$ parallel to the slope.

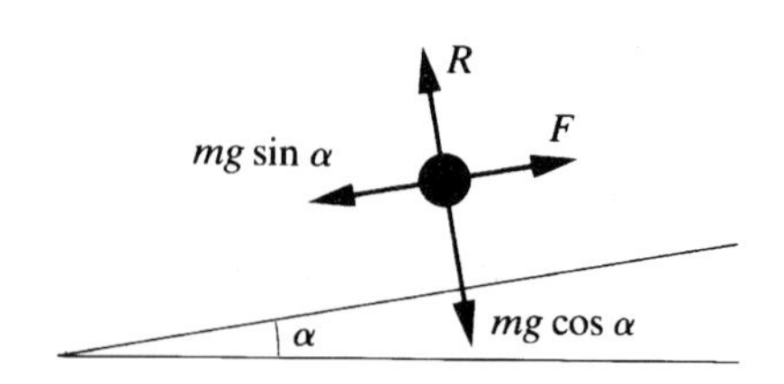

Since the skier is in equilibrium (at rest or moving with constant speed) applying Newton's second law:

Parallel to slope: $mg \sin \alpha - F = 0 \Rightarrow F = mg \sin \alpha$

Perpendicular to slope $R - mg \cos \alpha = 0 \Rightarrow R = mg \cos \alpha$

At rest on the point of moving or moving at constant speed,

$$F = \mu R$$

Substituting for F and R gives

$$mg \sin \alpha = \mu \, mg \cos \alpha$$

$$\Rightarrow \qquad \mu = \frac{\sin\alpha}{\cos\alpha} = \tan\alpha.$$

In this case $\mu = 0.35$, so $\tan \alpha = 0.35$

$$\alpha = 19.3°$$

NOTE

1. The result is independent of the mass of the skier. This is often found in simple mechanics models. For example, two objects of different mass fall to the ground with the same acceleration, the motion being independent of the masses of the objects. However when such models are refined, for example to take account of air resistance, mass is often found to have some effect on the result.

2. The angle for which the skier is about to slide down the slope is called the angle of friction. The angle of friction is often denoted by λ (lambda) and defined by $\tan \lambda = \mu$.

When the angle of the slope (α) is equal to the angle of the friction (λ), it is just possible for the skier to stand on the slope without sliding. If the slope is slightly steeper, the skier will slide immediately, and if it is less steep he or she will find it difficult to slide at all without using the ski poles.

Exercise 1A

1. A block of mass 10 kg is resting on a horizontal surface. It is being pulled by a horizontal force T (in Newtons), and is on the point of sliding. Find the coefficient of friction when

(i) $T = 9.8$ (ii) $T = 49$.

2. In each of the following situations state whether the block moves, and if so, the magnitude of its acceleration.

Exercise 1A continued

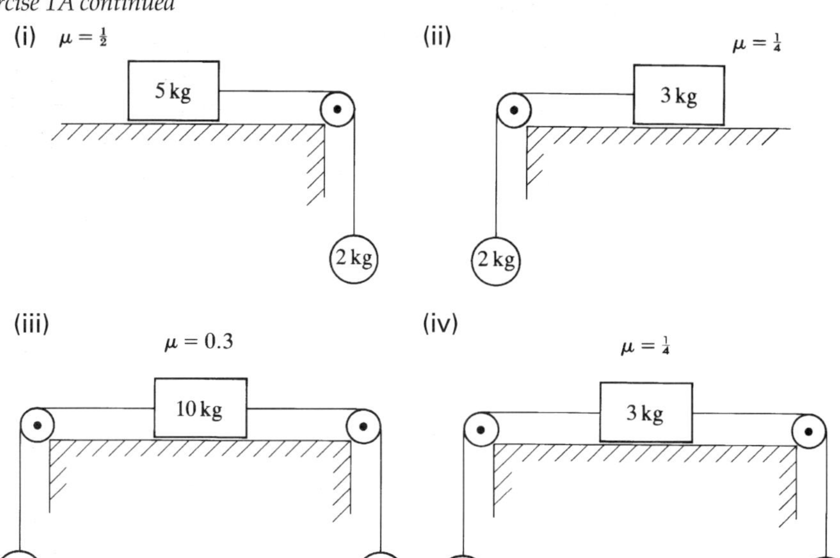

3. The brakes on a caravan of mass 700 kg have seized so that the wheels will not turn. What force must be exerted on the caravan to make it move horizontally? (The coefficient of friction between the tyres and the road is 0.7).

4. A boy slides a piece of ice of mass 100 g across the surface of a frozen lake. Its initial speed is 10 ms^{-1} and it takes 49 m to come to rest.

(i) Find the deceleration of the piece of ice.

(ii) Find the frictional force acting on the piece of ice.

(iii) Find the coefficient of friction between the piece of ice and the surface of the lake.

(iv) How far will a 200 g piece of ice travel if it, too, is given an initial speed of 10 ms^{-1}?

5. Jasmine is cycling at 12 ms^{-1} when she falls off. She slides a distance of 9 m before coming to rest. Calculate the coefficient of friction between Jasmine and the road.

6. A box of mass 50 kg is being moved across a room. To help it to slide a suitable mat is placed underneath the box.

(i) Explain why the mat makes it easier to slide the box.

A force of 98 N is needed to slide the box at a constant velocity.

(ii) What is the value of the coefficient of friction between the box and the floor?

A child of mass 20 kg climbs onto the box.

(iii) What force is now needed to slide the mat at constant velocity?

7. A car of mass 1200 kg is travelling at 20 ms^{-1} when it is forced to perform an emergency stop. Its wheels lock as soon as the brakes are applied so that they slide along the road without rotating. For the first 40 m the coefficient of friction between the wheels and the road is 0.75 but then the road surface changes and the coefficient of friction becomes 0.8.

(i) Find the deceleration of the car immediately after the brakes are applied.

(ii) Find the speed of the car when it comes to the change of road surface.

(iii) Find the total distance the car travels before it comes to rest.

8. Shona, whose mass is 30 kg, is sitting on a sledge of mass 10 kg which is being pulled at constant speed by her older brother, Aloke. The coefficient of friction between the sledge and the snow-covered ground is 0.15. Find the tension in the rope from Aloke's hand to the sledge when

(i) the rope is horizontal;

(ii) the rope makes an angle of 30° with the horizontal.

9. In each of the following situations a brick is about to slide down a rough inclined plane. Find the unknown quantity.

(i) The plane is inclined at 30° to the horizontal and the brick has mass 2 kg: find μ .

(ii) The brick has mass 4 kg and the coefficient of friction is 0.7: find the angle of the slope.

(iii) The plane is at 65° to the horizontal and the brick has mass 5 kg: find μ .

(iv) The brick has mass 6 kg and μ is 1.2: find the angle of slope.

10. The diagram shows a boy on a simple playground slide. The coefficient of friction between a typically clothed child and the slide is 0.25 and it can be assumed that no speed is lost when changing direction at B. The section AB is 3 m long and makes an angle of 40° with the horizontal. The slide is designed so that a child stops at just the right moment of arrival at C.

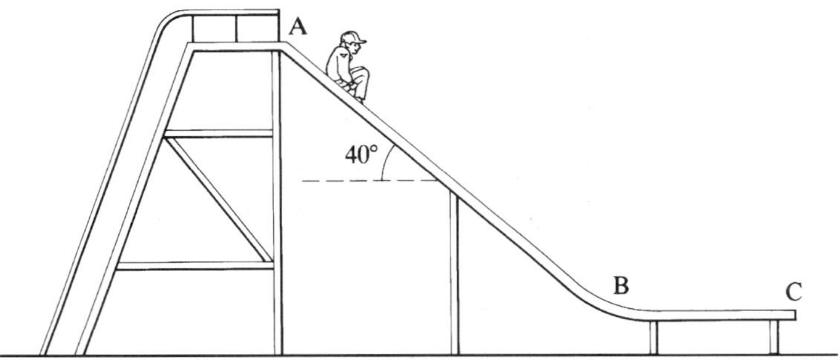

Exercise 1A continued

(i) Draw a diagram showing the forces acting on the boy when on the sloping section AB.

(ii) Calculate the acceleration of the boy when on the section AB.

(iii) Calculate the speed of the boy at B assuming that he started to slide from rest.

(iv) Find the length of the horizontal section BC.

11. The coefficient of friction between the skis and an artificial ski slope for learners is 0.3. During a run the angle, α, which the slope makes with the horizontal varies so that initially the skier accelerates, then travels at constant speed and then slows down. What can you say about the values of α in each of these three parts of the run?

12. The diagram shows a mop being used to clean the floor. The coefficient of friction between the mop and the floor is 0.3.

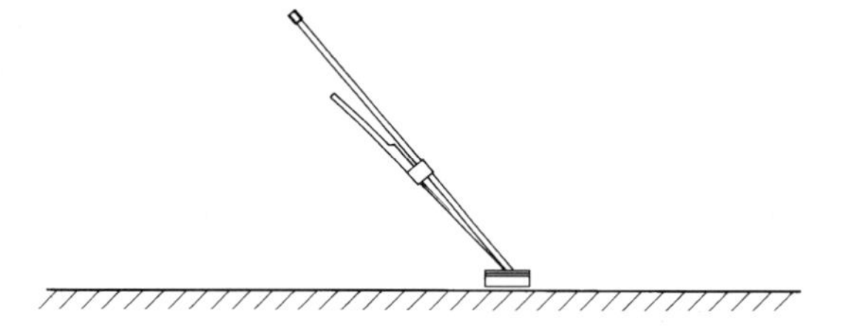

(i) Draw a diagram showing the forces acting on the head of the mop.

In an initial model the weight of the mop is assumed to be negligible.

(ii) Find the angle between the handle and the horizontal when the mop head is moving across the floor at constant velocity. Explain briefly why this angle is independent of how much force is exerted on the mop.

In a more refined model the weight of the head of the mop is taken to be $5\,\text{N}$, but the weight of the handle is still ignored.

(iii) Use this model to calculate the thrust in the handle if it is held at 70° to the horizontal while the head moves at constant velocity across the floor.

(iv) Could the same model be applied to a carpet cleaner on wheels? Explain your reasoning.

13. The diagram shows a block of mass $0.5\,\text{kg}$ resting on a rough inclined plane. The block is attached to a fixed point by a stretched elastic string, parallel to the plane. The coefficient of friction is 0.7 and the angle which the plane makes with the horizontal is given by $\alpha = \arcsin 0.6$.

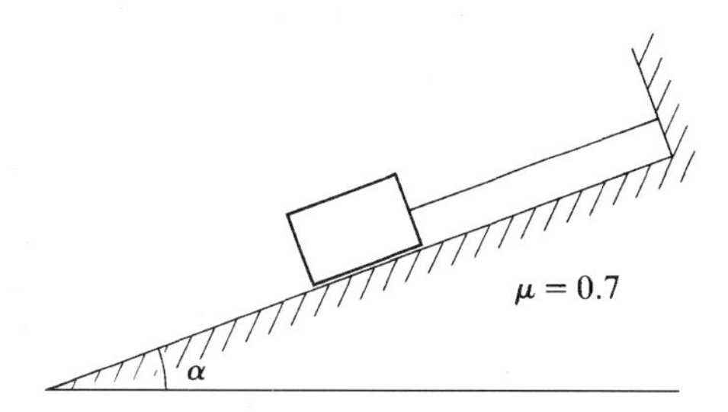

(i) Find the tension in the string when the block is on the point of sliding up the plane.

The block is pulled further down the plane from this position so that the tension is greater. It is then released and it slides up the plane for some distance before coming momentarily to rest with the string slack.

(ii) What happens next?

14. A shute at a water sports centre has been designed so that swimmers first slide down a steep part which is 10 m long and at an angle of 40° to the horizontal. They then come to a 20 m section with a gentler slope, 11° to the horizontal, where they travel at constant speed.

(i) Find the coefficient of friction between a swimmer and the shute.

(ii) Find the acceleration of a swimmer on the steep part.

(iii) Find the speed of a swimmer at the end of the shute. (You may assume that no speed is lost at the point where the slope changes).

An alternative design of shute has the same starting and finishing points but has a constant gradient.

(iv) With what speed do swimmers arrive at the end of this shute?

15. A box of weight 100 N is pulled at steady speed across a rough horizontal surface by a rope which makes an angle α with the horizontal. The coefficient of friction between the box and the surface is 0.4. Assume that the box slides on its underside and does not tip up.

(i) Find the tension in the string when the value of α is

(a) $10°$ (b) $20°$ (c) $30°$

(ii) Find an expression for the value of T for any angle α.

(iii) For what value of α is T a minimum?

Experiment

How good is Coulomb's model of friction?

Set up the apparatus shown below and use it in two ways:

(a) to find the value of μ when the slider is on the point of moving;

(b) to find the value of μ when the slider is moving.

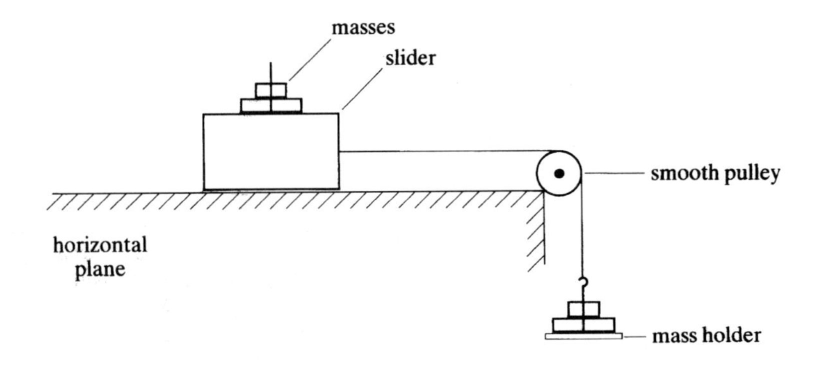

(a) Static Friction

Add small masses such as 5 g to the mass holder until the slider just begins to move. When adding the masses to the mass holder do so very gently and do not drop them onto it. Repeat the experiment several times and record all your results.

Find the greatest mass on the hanging mass holder for which the system does not move. From a knowledge of this mass, and the mass of the slider you can deduce the value of μ.

m_1 (kg)	$R = m_1g$ (N)	m_2 (kg)	$T = m_2g$ (N)
0.065	0.64	0.03	0.29
0.065	0.64	0.035	0.34
0.065	0.64	0.025	0.245
...	...	...	...

Now add various masses to the slider, say 100 g, 200 g, etc. and repeat the experiment for each of them. Plot all your results on a graph of T (vertical axis) against R (horizontal axis).

What can you conclude about the coefficient of friction, μ?

(b) Sliding Friction

Now repeat the experiment but this time place sufficiently large masses on the mass holder to ensure that the slider accelerates across the table. Start the slider at rest and time how long it takes to travel a fixed distance across the table. As before, take a number of measurements for various different masses on the slider and the holder.

From your measurements deduce the force of resistance acting on the slider, and so the coefficient of friction.

Is the value of the coefficient of friction

(i) the same for different values of the masses m_1 and m_2?

(ii) the same for static friction as for sliding friction?

Investigations

The sliding ruler

Hold a metre ruler horizontally across your two index fingers and slide your fingers smoothly together, fairly slowly. What happens?

Use the laws of friction to investigate what you observe.

Optimum Angle

A packing case is pulled across rough ground by means of a rope making an angle θ with the horizontal. Investigate how the tension can be minimised by varying the angles of the rope to the horizontal.

Life without friction

Friction forces are essential in many real situations, so life without friction might be rather difficult. Could you survive without friction?

Swimming Pool Flume

Flumes are a popular attraction at many modern swimming pools. Observe the motion of someone on a swimming pool flume, and deduce the value of the coefficient of friction between the person and the flume.

Can Coulomb's model be applied to this situation?

KEY POINTS

Coulomb's Law

- The frictional force, F, between two surfaces is given by

$$F \leq \mu R \text{ when there is no sliding}$$
$$F = \mu R \text{ when sliding occurs}$$

where R is the normal reaction of one surface on the other and μ is the coefficient of friction between the surfaces.

- The frictional force always acts in the direction to oppose sliding.

2 Moments of forces

Give me a firm place to stand and I will move the earth.

Archimedes

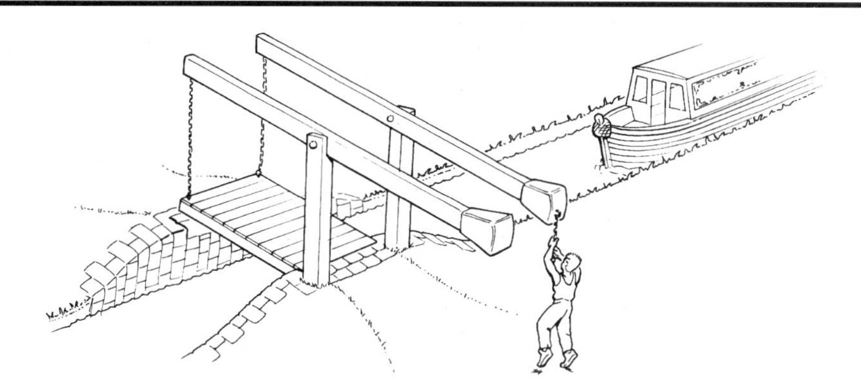

The illustration shows a swing bridge over a canal. It can be raised to allow barges and boats to pass. It is operated by hand, even though it is very heavy. How is this possible?

The bridge depends on the turning effects or *moments* of forces. To understand these you may find it helpful to look at a simpler situation.

Two children sit on a simple see-saw, made of a plank balanced on a fulcrum as in figure 2.1. Will the see-saw balance?

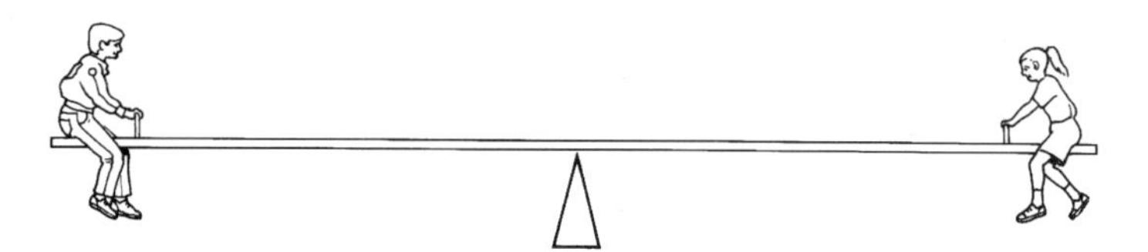

Figure 2.1

If both children have the same mass and sit the same distance from the fulcrum, then you expect the see-saw to balance.

Now consider possible changes to this situation:

(i) If one child is heavier than the other then you expect the heavier one to go down;

(ii) If one child moves nearer the centre you expect that child to go up.

You can see that both the weights of the children and their distances from the fulcrum are important.

What about this case? One child has mass 35 kg and sits 1.6 m from the fulcrum and the other has mass 40 kg and sits on the opposite side 1.4 m from the fulcrum (see figure 2.2).

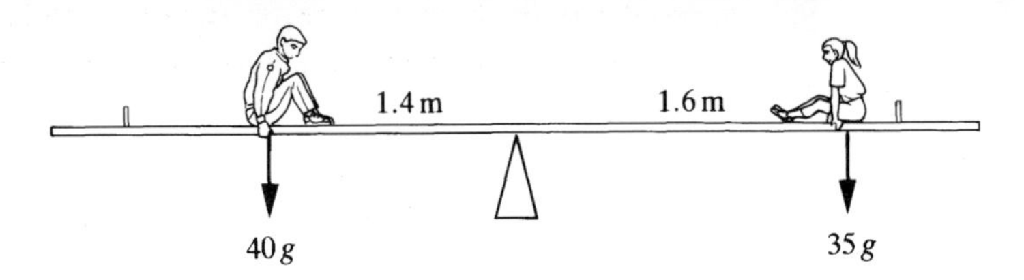

Figure 2.2

Taking the products of their weights and their distances from the fulcrum, gives

$$\text{A: } 40\,g \times 1.4 = 56\,g$$

$$\text{B: } 35\,g \times 1.6 = 56\,g$$

So you might expect the see-saw to balance and this indeed is what would happen.

Rigid bodies

Until now the particle model has provided a reasonable basis for the analysis of the situations you have met. In examples like the see-saw however, where turning is important, this model is inadequate because the forces do not all act through the same point.

In such cases you need the *rigid body model* in which the body is recognised as having size and shape, but is assumed not to be deformed when forces act on it.

Suppose that you push a tray with one finger so that the force acts parallel to one edge and through the centre of mass (figure 2.3).

Figure 2.3

The particle model is adequate here: the tray travels in a straight line in the direction of the applied force.

If you push the tray equally hard with two fingers as in figure 2.4, symmetrically either side of the centre of mass, the particle model is still adequate.

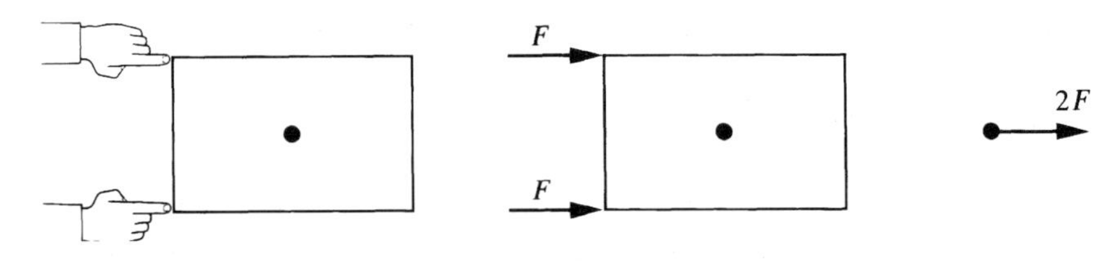

Figure 2.4

However if the two forces are not equal or are not symmetrically placed, or as in figure 2.5 are in different directions, the particle model cannot be used.

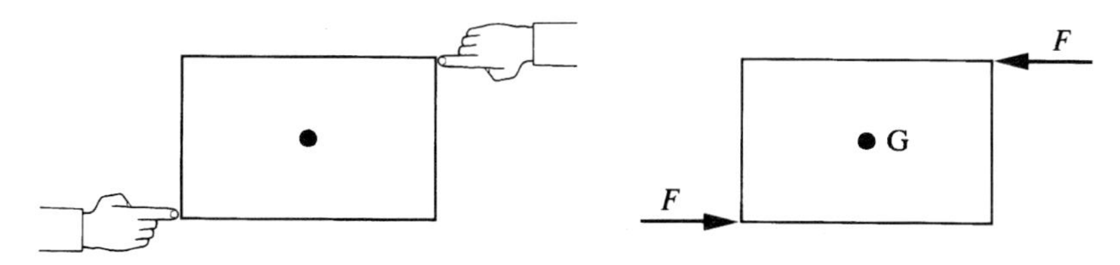

Figure 2.5

The resultant force is now zero, since the individual forces are equal in magnitude but opposite in direction. What happens to the tray? Experience tells us that it starts to rotate. How fast it starts to rotate depends, among other things, on the magnitude of the forces and the width of the tray. The rigid body model allows you to analyse the situation.

Moments

In the example of the see-saw we looked at the product of each force and its distance from a fixed point. This product is called the *moment* of the force about the point.

The see-saw balances because the moments of the forces on either side of the fulcrum are the same magnitude and in opposite directions. One would tend to make the see-saw turn clockwise, the other anti-clockwise. By contrast, the moments of the forces on the tray in the last situation do not balance. They both tend to turn it anticlockwise, so rotation occurs.

Conventions and units

The moment of a force F about a point O is defined by

$$\text{moment} = Fd$$

where d is the perpendicular distance from the point O to the line of action of the force (figure 2.6).

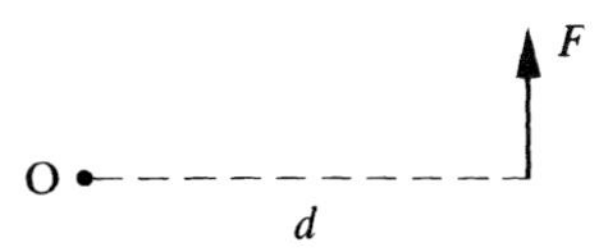

Figure 2.6

In two dimensions, the sense of a moment is described as either positive (anticlockwise) or negative (clockwise) as shown in figure 2.7.

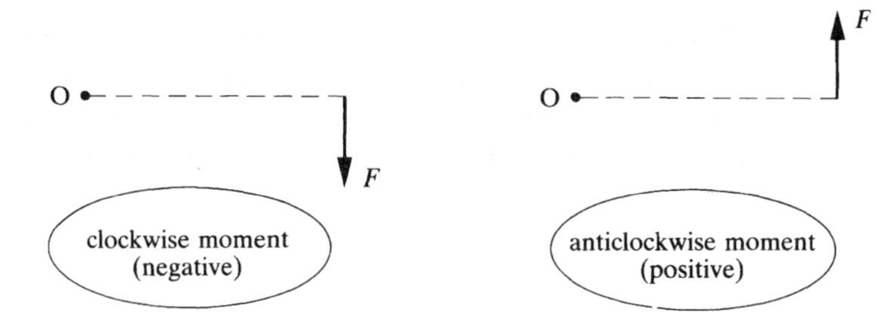

Figure 2.7

In the SI system the unit for moment is the newton metre (Nm), because a moment is the product of a force, the unit of which is the newton, and distance, the unit of which is the metre.

Remember that moments are always taken about a point and you must always specify what that point is. A force acting through the point will have no moment about that point because in that case $d = 0$.

For Discussion

The photographs show two tools for undoing wheel nuts on a car. Discuss the advantages and disadvantages of each.

When using the spider wrench (the tool with two 'arms'), you apply equal and opposite forces either side of the nut. These produce moments in the same direction. One advantage of this method is that there is no resultant force and hence no tendency for the nut to snap off.

Couples

Whenever two forces of the same magnitude act in opposite directions along different lines, they have a zero resultant force, but do have a turning effect. In fact the moment will be Fd about any point where d is the perpendicular distance between the forces. This is demonstrated in figure 2.8.

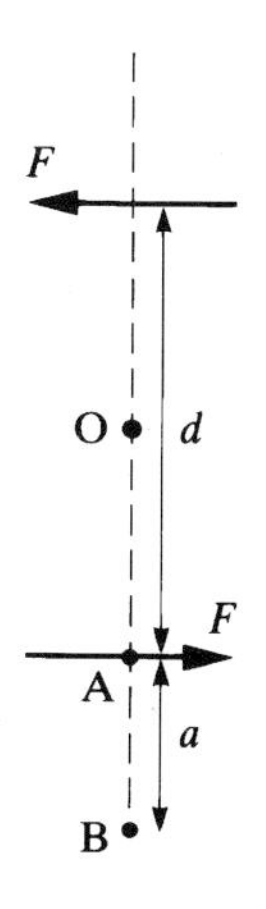

Figure 2.8

Moment about O is $F\frac{d}{2} + F\frac{d}{2} = Fd$

Moment about A is $0 + Fd = Fd$

Moment about B is $-aF + (a+d)F = Fd$

A pair of forces like this with a zero resultant but a non-zero total moment is known as a couple. The effect of a couple on a rigid body is to cause rotation.

Equilibrium revisited

In *Mechanics 1* we said that an object is in equilibrium if the resultant force on the object is zero. This definition is adequate provided all the forces act through the same point on the object. However we are now concerned with forces acting at different points, and in this situation even if the forces balance there may be a resultant moment.

Figure 2.9 shows a tray being pushed equally hard at opposite corners.

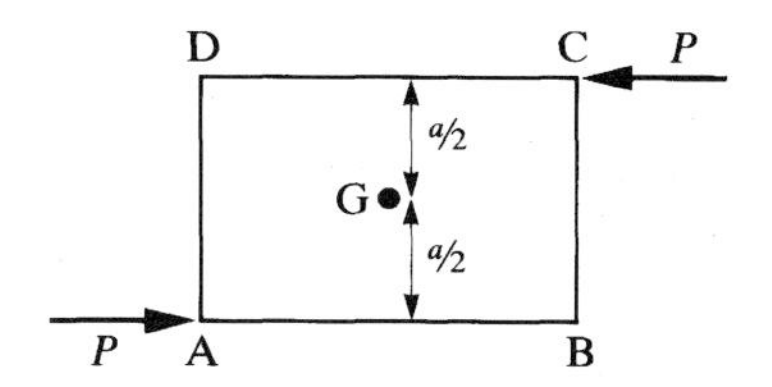

Figure 2.9

The resultant force on the tray is clearly zero, but the resultant moment about its centre point, G, is

$$P \times \frac{a}{2} + P \times \frac{a}{2} = Pa.$$

The tray will start to rotate about its centre and so it is clearly not in equilibrium.

NOTE

You could have taken moments about any of the corners, A, B, C or D, or any other point in the plane of the paper and the answer would have been the same, Pa anticlockwise.

So we now tighten our mathematical definition of equilibrium to include moments. For an object to remain at rest (or moving at constant velocity) when a system of forces is applied, both the resultant force and the total moment must be zero.

To check that an object is in equilibrium under the action of a system of forces, you need to check two things:
(i) that the resultant force is zero;
(ii) that the resultant moment about any point is zero. (You only need to check one point).

For Discussion

Look again at the swing bridge on page 17. It is in equilibrium, both in the open position, and in the closed position.

(i) What would happen if someone raised the bridge to 45° and then left?

(ii) The illustration shows an adult operating the bridge. Could a small, lightweight child operate it successfully?

EXAMPLE

Two children are playing with a door. Kerry tries to open it by pulling on the handle with a force 50 N at right angles to the plane of the door, at a distance 0.8 m from the hinges. Peter pushes at a point 0.6 m from the hinges, also at right angles to the door and with sufficient force just to stop Kerry opening it.
(i) What is the moment of Kerry's force about the hinges?
(ii) With what force does Peter push?
(iii) Describe the resultant force on the hinges.

Solution

Looking down from above, the door opens clockwise. Anticlockwise is taken to be positive.

(i)

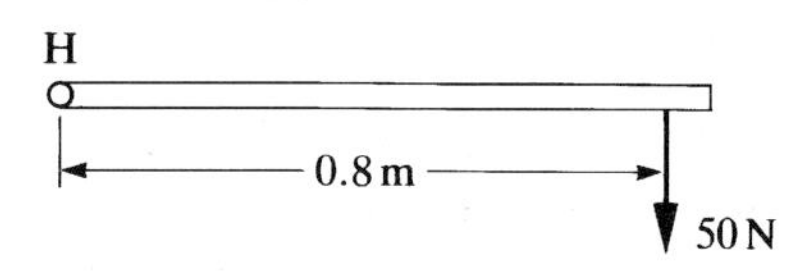

Kerry's moment about H $= -50 \times 0.8$

$= -40$ Nm

The moment of Kerry's force about the hinges is -40 Nm.
(Note that it is a clockwise moment and so negative).

(ii)

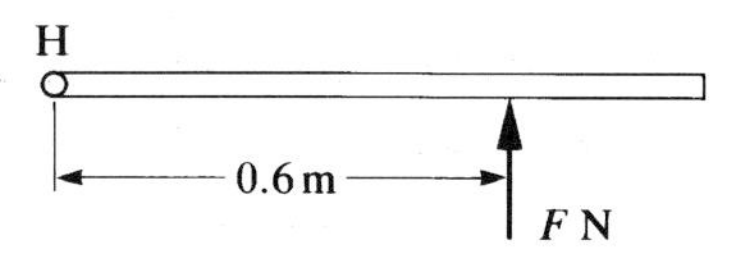

Peter's moment about H $= +F \times 0.6$

Since the door is in equilibrium, the total moment on it must be zero, so

$$F \times 0.6 - 40 = 0$$

$$F = \frac{40}{0.6}$$

$$= 66\tfrac{2}{3}$$

Peter pushes with a force of $66\frac{2}{3}$ N.

(iii) Since the door is in equilibrium the overall resultant force on it must be zero.

All the forces are at right angles to the door, as shown in the diagram.

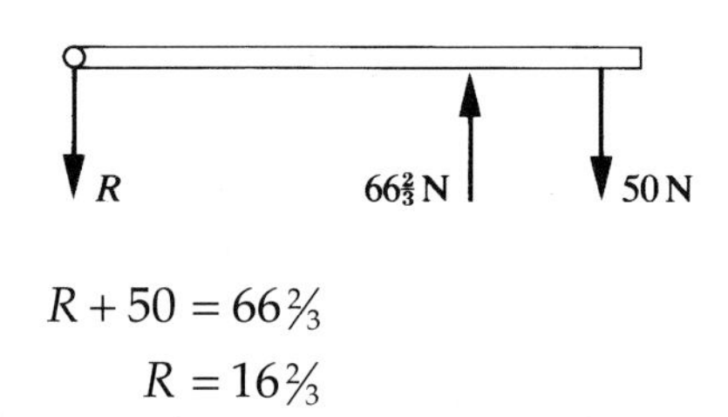

$$R + 50 = 66\tfrac{2}{3}$$
$$R = 16\tfrac{2}{3}$$

The reaction at the hinge is a force of $16\tfrac{2}{3}$ N in the same direction as Kerry is pulling.

NOTE

The reaction force at a hinge may act in any direction, according to the forces elsewhere in the system. A hinge can be visualised in cross section as shown in figure 2.10. If the hinge is well oiled, and the friction between the inner and outer parts is negligible, the hinge cannot exert any moment. In this situation the door is said to be 'freely hinged'.

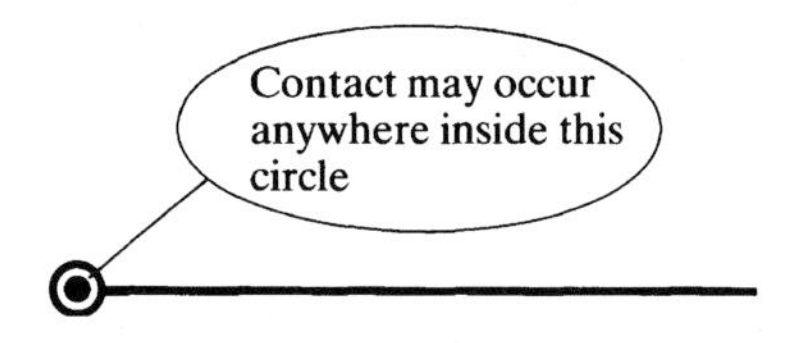

Figure 2.10

EXAMPLE

The diagram shows a man of weight 600 N standing on a footbridge that consists of a uniform wooden plank just over 2 m long of weight 200 N. Find the reaction forces exerted on each end of the plank.

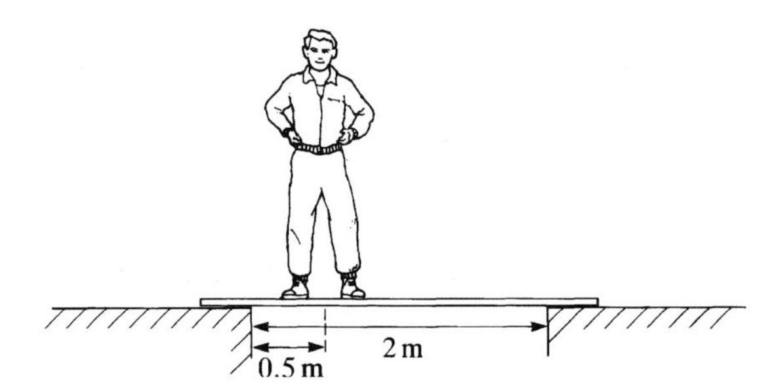

Solution

The diagram shows the forces acting on the plank.

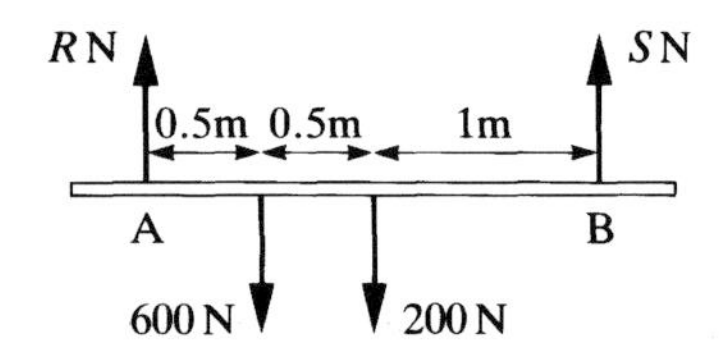

For equilibrium both the resultant force and the total moment must be zero.

As all the forces act vertically we have

$$R + S - 800 = 0 \qquad (1)$$

Taking moments about the point A gives

$$R \times 0 - 600 \times 0.5 - 200 \times 1 + S \times 2 = 0 \qquad (2)$$

From equation (2) $S = 250$ and so equation (1) gives $R = 550$

The reaction forces are 250 N at A and 550 N at B

NOTE

1. *You cannot solve this problem without taking moments.*
2. *You can take moments about any point and can, for example, show that by taking moments about B you get the same answer.*
3. *The whole mass of the plank is being considered to act at its centre.*
4. *The moment of a force through the point about which moments are being taken is zero.*

Levers

A lever is used to lift or move a heavy object using a relatively small force. Levers depend on moments for their action.

The two most common lever configurations are shown below. In both cases a load W is being lifted by an applied force F, using a lever of length l. The calculations assume equilibrium.

Case 1
The fulcrum is at one end of the lever, figure 2.11.

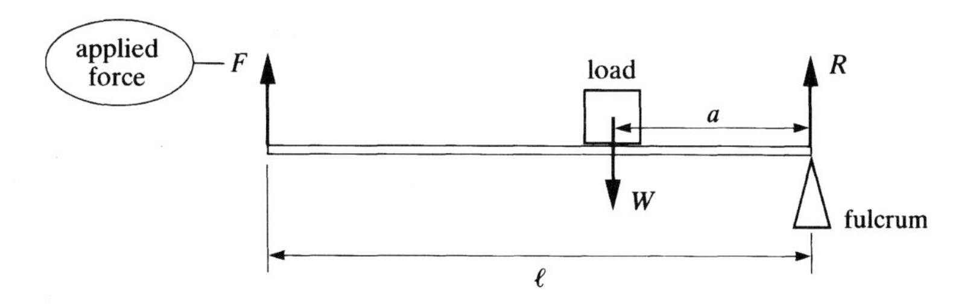

Figure 2.11

Taking moments about the fulcrum

$$F \times l - W \times a = 0$$

$$F = W \times \frac{a}{l}$$

Since a is much smaller than l, the applied force F is much smaller than the load W.

Case 2

The fulcrum is within the lever, figure 2.12.

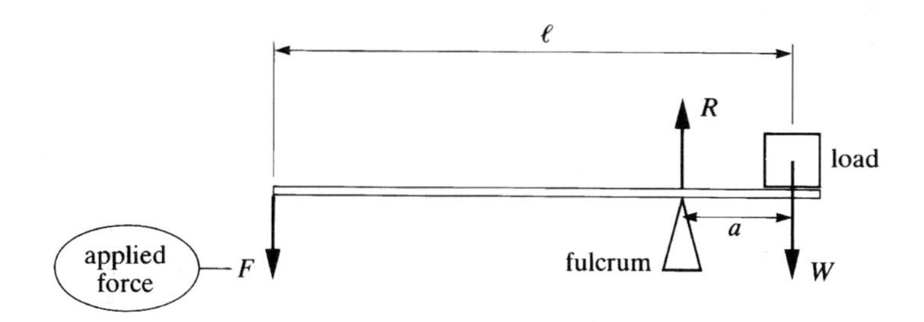

Figure 2.12

Taking moments about the fulcrum

$$F \times (l - a) - W \times a = 0$$

$$F = W \frac{a}{l - a}$$

Provided that the fulcrum is nearer the end with the load, the applied force is less than the load.

For Discussion

Under what circumstances would you use each configuration?

Exercise 2A

1. In each of the situations shown below, find the moment of the force about the point and state whether it is positive (anticlockwise) or negative (clockwise).

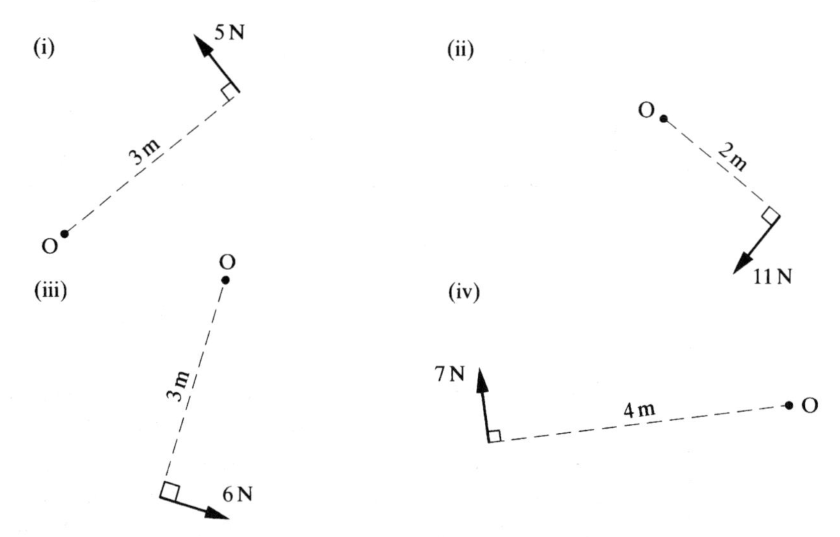

2. The situations below involve several forces acting on each object. For each one, find the total moment.

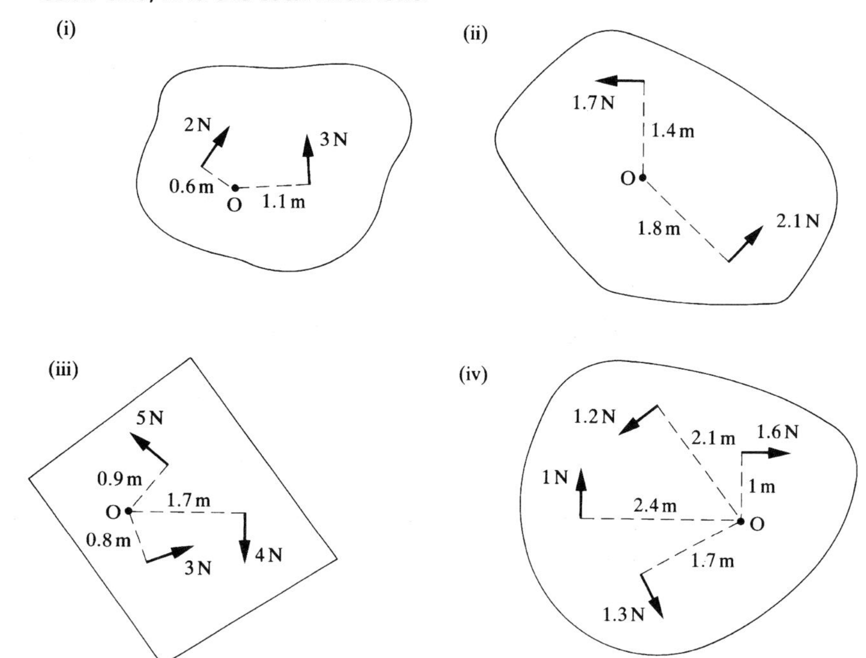

3. A uniform horizontal bar of mass 5 kg has length 30 cm and rests on two vertical supports, 10 cm and 22 cm from its left hand end. Find the magnitude of the reaction force at each of the supports.

4. Find the reaction forces on the hi-fi shelf shown below. The shelf itself has weight 25 N and its centre of mass is midway between A and D.

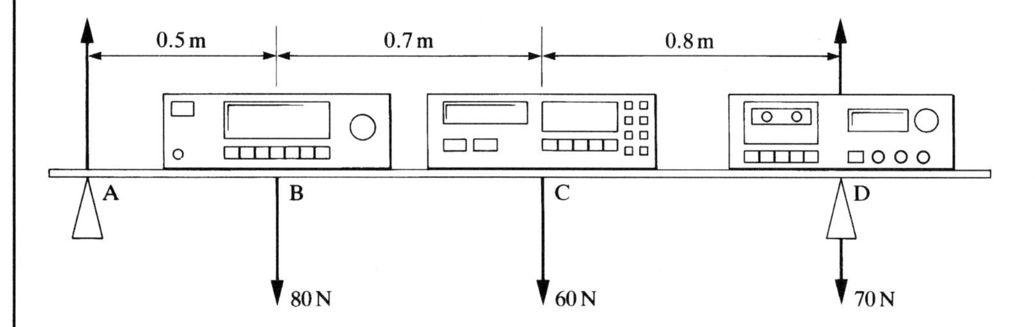

5. The diagram shows a motorcycle of mass 250 kg, and its rider whose mass is 80 kg. The centre of mass of the motorcycle lies on a vertical line midway between its wheels. When the rider is on the motorcycle, his centre of mass is 1 m behind the front wheel. Find the vertical reaction forces acting through the front and rear wheels when

Exercise 2A continued

(i) the rider is not on the motorcycle,

(ii) the rider is on the motorcycle.

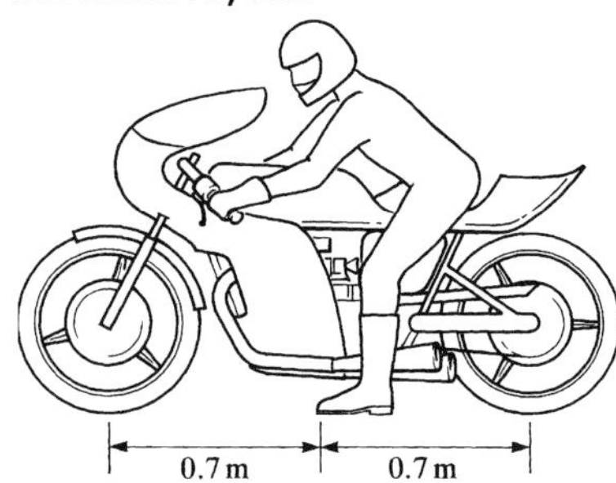

6. The diagram shows a uniform roof truss of weight 400 N, which is carrying loads of 300 N, 700 N, and 200 N. Find the reaction forces, R_1 and R_2 in Newtons, exerted on the truss by its supports.

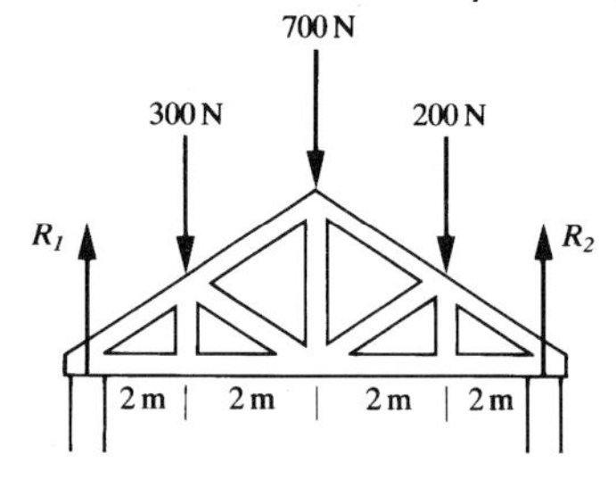

7. The diagram shows two people, an adult and a child, sitting on a uniform bench of mass 40 kg; their positions are as shown. The mass of the child is 50 kg, that of the adult is 85 kg.

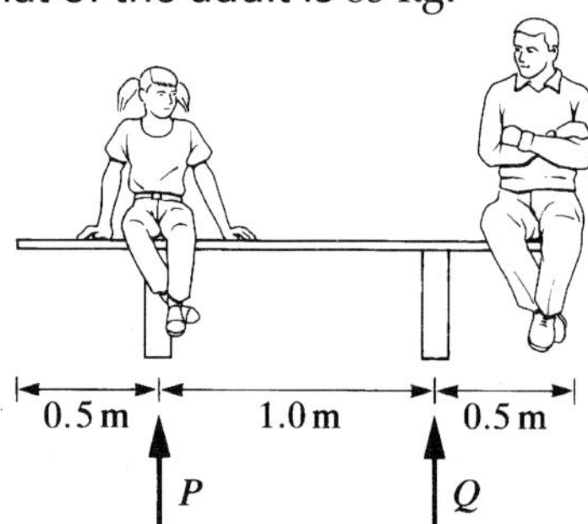

(i) Find the reaction forces, P and Q in Newtons, from the ground on the two supports of the bench.

(ii) The child now moves to the mid-point of the bench. What are the new values of P and Q ?

(iii) Is it possible for the child to move to a position where $P = 0$? What is the significance of a zero value for P?

(iv) What happens if the child leaves the bench?

8. The diagram shows a diving board which some children have made. It consists of a uniform plank of mass 20 kg and length 3 m, with 1 m of its length projecting out over a canal. They have placed a stone of mass

25 kg above the support at the end over the land; there is another support at the water's edge.

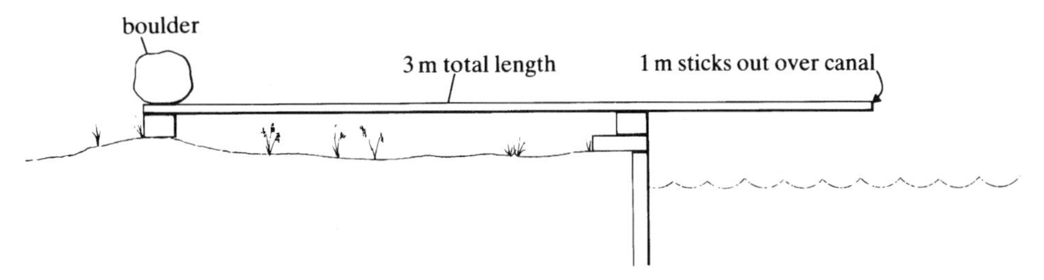

(i) Find the forces at the two supports when no one is using the diving board.

(ii) A child of mass 50 kg is standing on the end of the diving board over the canal. What are the forces at the two supports?

(iii) Some older children arrive and take over the diving board. One of these is a heavy boy of mass 90 kg. Describe as precisely as you can what happens when he uses the diving board.

9. A lorry of mass 5000 kg is driven across a Bailey bridge of mass 20 tonnes. The length of the bridge is 10 m.

(i) Find expressions for the reaction forces at each end of the bridge in terms of the distance x in metres travelled by the lorry from the start of the bridge.

(ii) From what point of the lorry is the distance x measured?

Two identical lorries cross the bridge at the same speed, starting at the same instant, from opposite directions.

(iii) How do the reaction forces of the supports on the bridge vary as the lorries cross the bridge?

10. A simple suspension bridge across a narrow river consists of a uniform beam, 4 m long and of mass 60 kg, supported by vertical cables attached at a distance 0.75 m from each end of the beam.

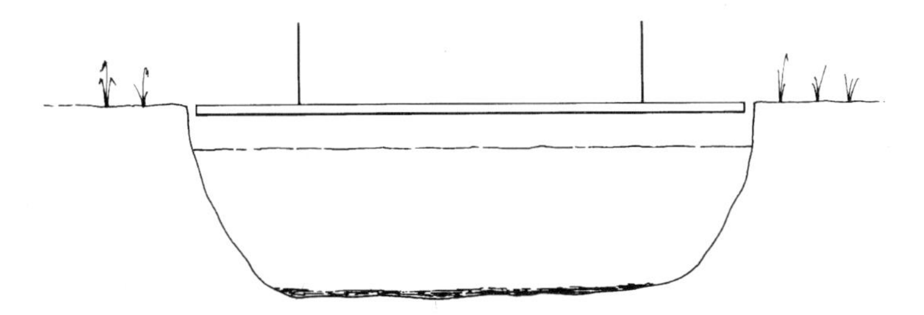

(i) Find the tension in each cable when a boy of mass 50 kg stands 1 m from the end of the bridge.

(ii) Can a couple walking hand-in-hand cross the bridge safely, without it tipping, if their combined mass is 115 kg?

Exercise 2A continued

(iii) What is the mass of a person standing on the end of the bridge when the tension in one cable is double that in the other cable?

11. The diagram shows a stone slab AB of mass 1 tonne resting on two supports, C and D. The stone is uniform and has length 3 m. The supports are at distances 1.2 m and 0.5 m from the end.

(i) Find the reaction forces at the two supports.

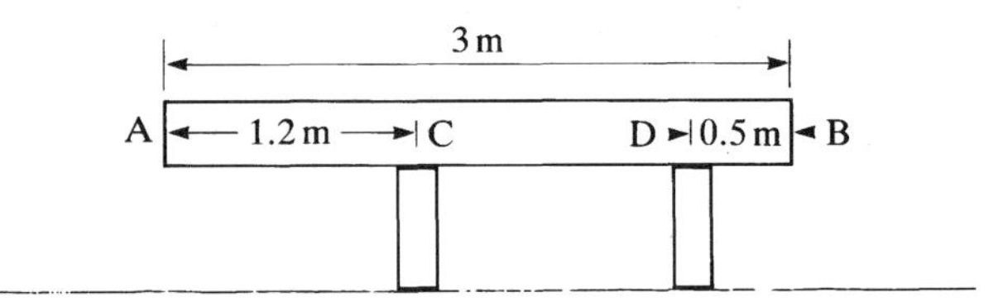

Local residents are worried that the arrangement is unsafe since their children play on the stone.

(ii) How many children each of mass 50 kg would need to stand at A in order to tip the stone over?

The stone's owner decides to move the support at C to a point nearer to A. To take the weight of the slab while doing this, he sets up the lever system shown in the diagram. The distance XF is 1.25 m and FY is 0.25 m.

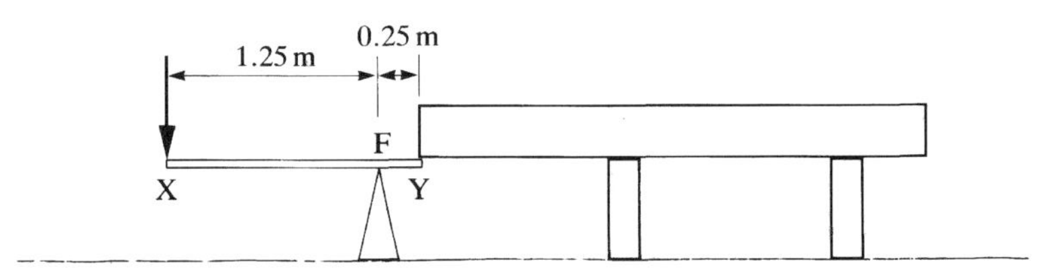

(iii) What downward force applied at X would reduce the reaction force at C to zero (and so allow the support to be moved)?

12. Four seamen are using a light capstan to pull in their ship's anchor – one of them is shown in the diagram. The diameter of the capstan's drum is 1 m and the spokes on which the men are pushing each project 2 m from the centre of the capstan. Each man is pushing with a force of 300 N, horizontally and at right angles to his spoke. The anchor cable is taut; it passes over a frictionless pulley and then makes an angle of 20° with the horizontal.

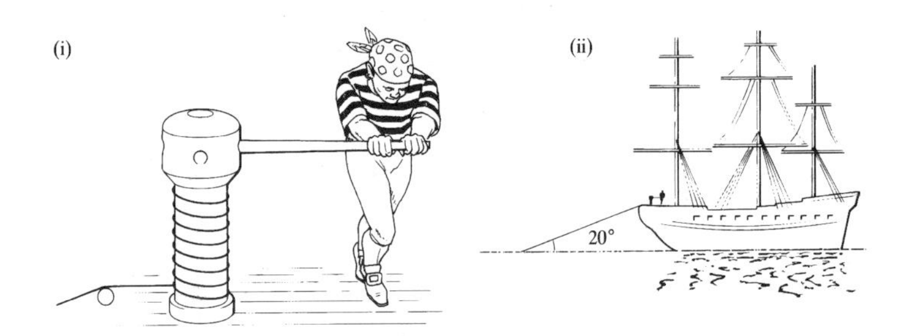

(i) Find the tension in the cable.

The mass of the ship is 2000 tonnes.

(ii) Find the acceleration of the ship, assuming that no other horizontal forces act on the ship.

In fact the acceleration of the ship is $0.0015 \mathrm{~ms}^{-2}$. Part of the difference can be explained by friction with the capstan, resulting in a resisting moment of 300 Nm, the rest by the force of resistance, R N, to the ship passing through the water.

(iii) Find the value of R.

Experiment

Set up the apparatus shown in the photograph below.

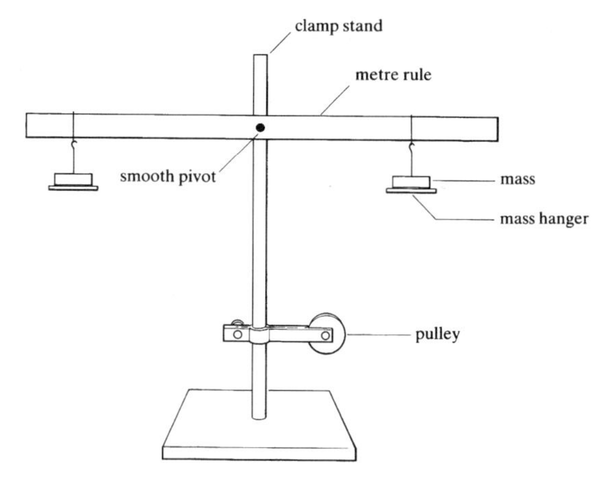

What forces act on the metre rule?

Try to predict the outcome of each experiment before you start to take any measurements.

1. A mass is placed at the left hand end of the metre rule. Can the rule be balanced by (a) a lighter mass to the right of the support or (b) a heavier mass to the right of the support?

2. A mass of 100 g is hung from the left hand end of the metre rule. Where will masses of 100 g, 200 g, 300 g and 400 g need to be hung if the rule is to balance?

3. Two masses are suspended from the rule in such a way that the rule balances in a horizontal position. The rule is then moved to an inclined position and released.

What happens?

4. A string is attached to one end of the rule and passed over a pulley as shown in the diagram. If the rule is to balance now, which mass must be greater?

Experiment continued

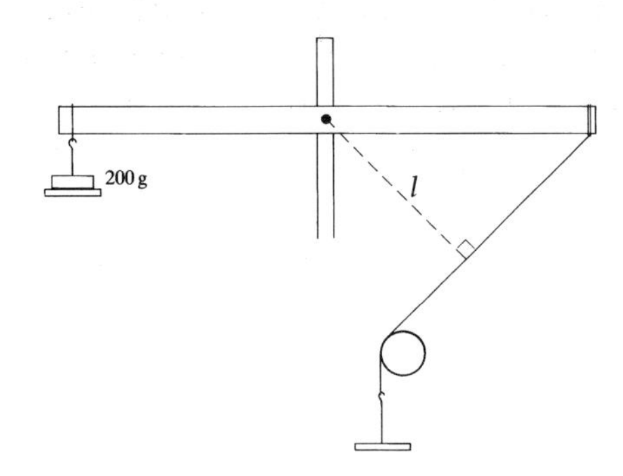

Measure the distance l. What do you notice?

Repeat this experiment with different masses and with different pulley positions.

The moment of a force which acts at an angle

From the experiment you will have seen that the moment of a force about the pivot depends on the *perpendicular distance* between the force and the pivot, rather than on the distance between the *point of application* of the force and the pivot.

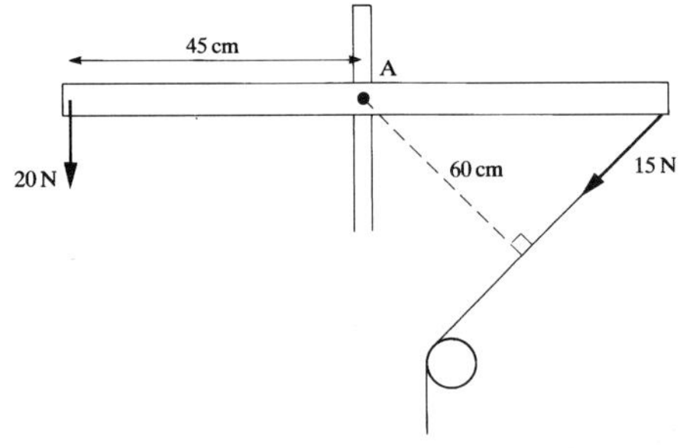

Figure 2.13

In figure 2.13, where the system remains at rest, the moment about A of the 20 N force is $20 \times 0.45 = 9$ Nm. The moment about A of the 15 N force is $-15 \times 0.6 = -9$ Nm. The system is in equilibrium even though unequal forces act at equal distances from the pivot.

The magnitude of the moment of the force F about O in figure 2.14 is given by

$$\begin{aligned} \text{Moment about O} &= F \times l \\ &= Fd \sin \alpha \end{aligned}$$

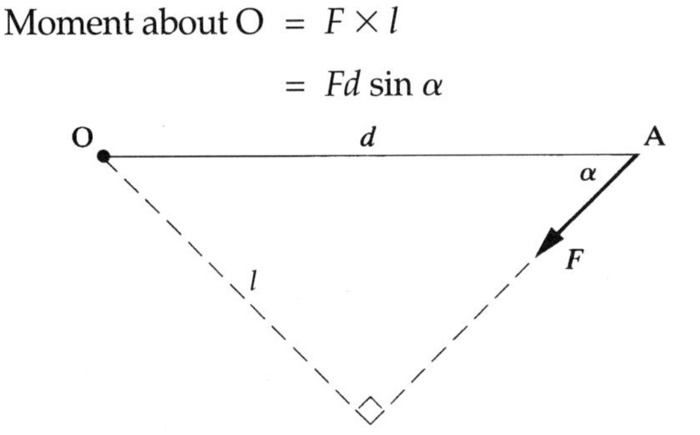

Figure 2.14

Alternatively the moment can be found by noting that the force F can be split into components $F\cos\alpha$ parallel to AO and $F\sin\alpha$ perpendicular to AO, both acting through A (figure 2.15). The moment of each component can be found and then summed to give the total moment.

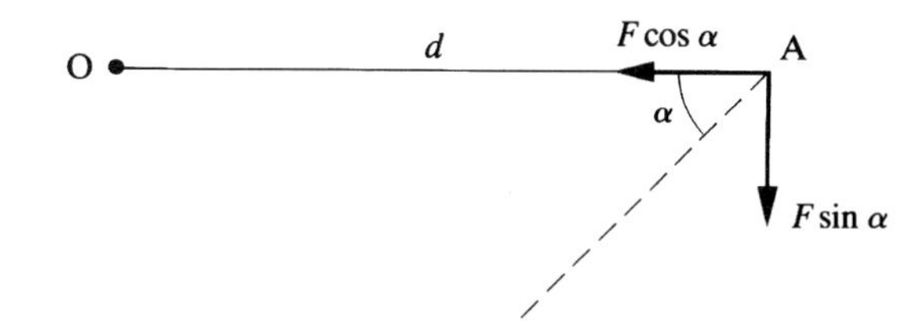

Figure 2.15

The moment of the component parallel to AO is zero because it acts through O. The moment of the perpendicular component is $Fd\sin\alpha$, so the total moment is $Fd\sin\alpha$, as expected.

EXAMPLE

A force of 40 N is exerted on a rod as shown. Find the moment of the force about the point marked O.

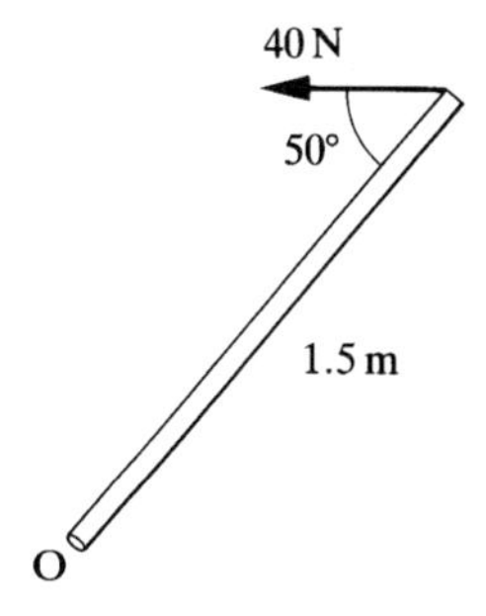

Solution

In order to calculate the moment, the perpendicular distance between O and the line of action of the force must be found. This is shown in the diagram.

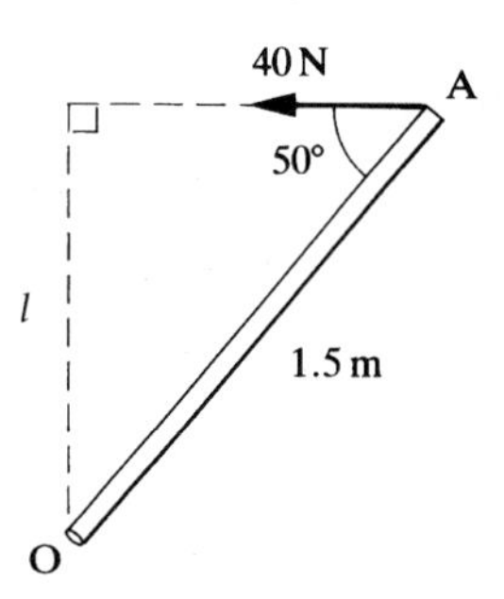

Here $l = 1.5 \times \sin 50°$.

So the moment about O is

$$F \times l = 40 \times (1.5 \times \sin 50°)$$
$$= 46.0 \text{ in Nm.}$$

Alternatively you can resolve the 40 N force into components as in the next diagram.

The component of the force parallel to AO is 40cos50° newtons. The component perpendicular to AO is 40cos40° (or 40sin50°) newtons.

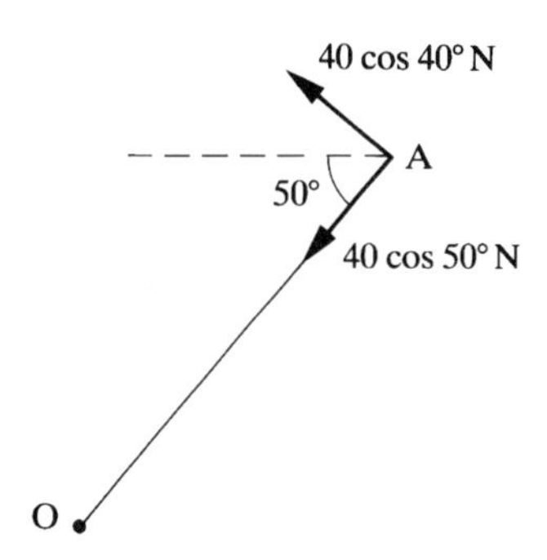

So the moment about O is $40\sin 50° \times 1.5 = 60\cos 40°$

$= 46.0$ Nm as before

EXAMPLE

A sign outside a pub is attached to a light rod of length 1 m which is freely hinged to the wall and supported in a vertical plane by a light string as in the diagram. The sign is assumed to be a uniform rectangle of mass 10 kg. The angle of the string to the horizontal is 25°.

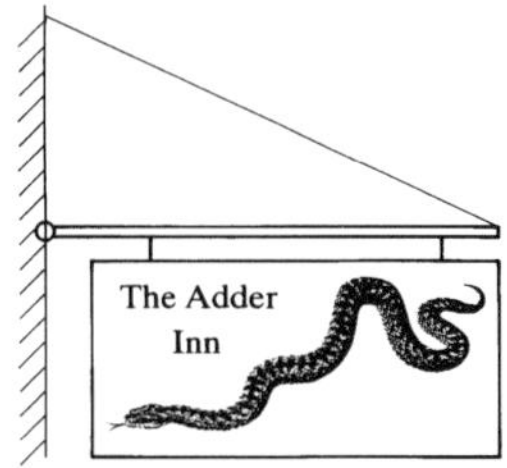

(i) Find the tension in the string.
(ii) Find the magnitude and direction of the reaction force of the hinge on the sign.

Solution

(i) The diagram shows the forces acting on the rod, where R_H and R_V are the magnitudes of the horizontal and vertical components of the reaction **R** on the rod at the wall.

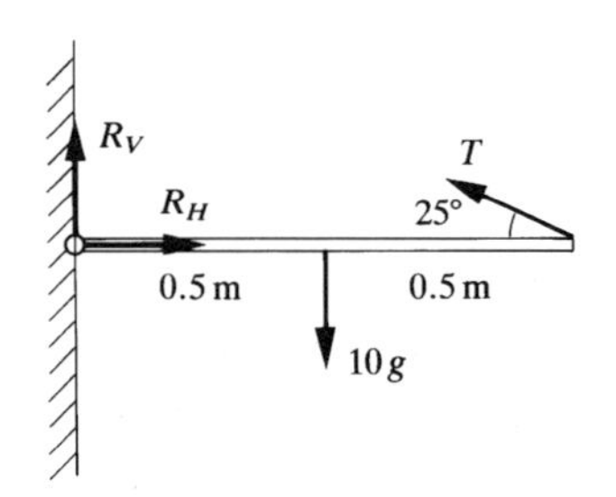

Taking moments about O

$$R \times 0 - 10g \times 5 + T\sin 25° \times 1 = 0$$

$$\Rightarrow \qquad \frac{5g}{\sin 25°} = 58.0$$

The tension is 58.0 N.

(ii) We can find the reaction at the wall by resolving.

Horizontally: $R_H = T\cos 25° \quad \Rightarrow R_H = 52.5$

Vertically: $R_V + T\sin 25° = 10g \quad \Rightarrow R_V = 73.5$

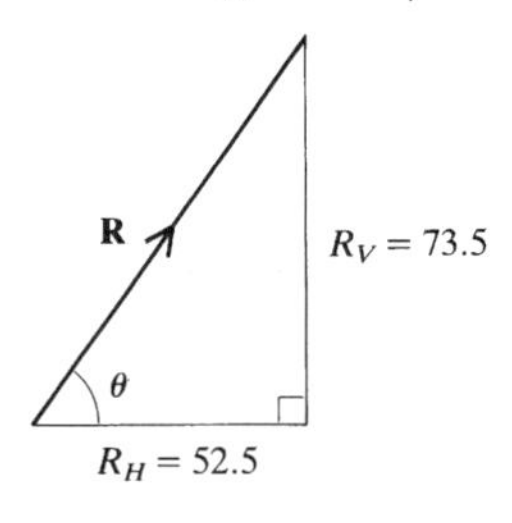

$$R = \sqrt{52.5^2 + 73.5^2} = 90.3$$

$$\theta = \arctan\left(\frac{73.5}{52.5}\right) = 54.5$$

The reaction at the hinge has magnitude 90.3 N and acts at 54.5° above the horizontal.

EXAMPLE

A ladder is standing on rough ground and leaning against a smooth wall at an angle of 60° to the ground. The ladder has length 4 m and mass 15 kg. Find the reaction force at the wall and ground and the friction force at the ground.

Solution

The diagram shows the forces acting on the ladder. The forces are in newtons.

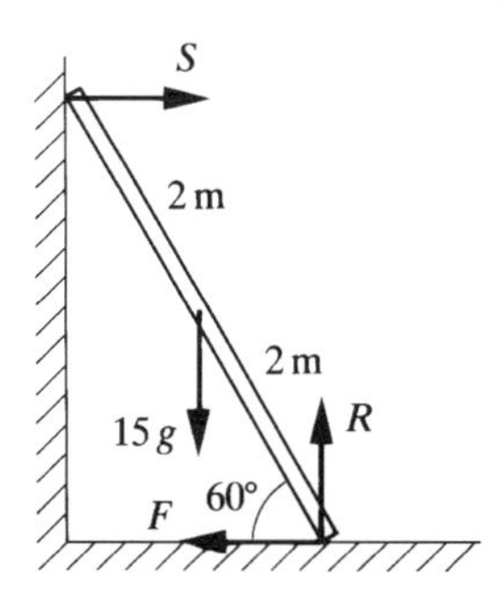

The diagram shows that there are three unknown forces S, R and F so we need three equations from which to find them. If the ladder remains at rest (in equilibrium) then the resultant force is zero and the resultant moment is zero. These two conditions provide the three necessary equations.

Equilibrium of horizontal components: $S - F = 0$

Equilibrium of vertical components: $R - 15g = 0$

Rotational equilibrium: $R \times 0 + F \times 0 + 15g\sin 30° \times 2 + S\sin 60° \times 4 = 0$

$$\Rightarrow \quad 147 + 4S\sin 60° = 0$$

Solving for R, S and F gives $R = 147$, $S = 42.4$ and $F = 42.4$, (taking g to be 9.8 in ms^{-2}).

The force at the wall is 42.4 N, those at the ground are 42.4 N horizontally and 147 N vertically.

Exercise 2B

1. Find the moment about O of each of the forces illustrated below.

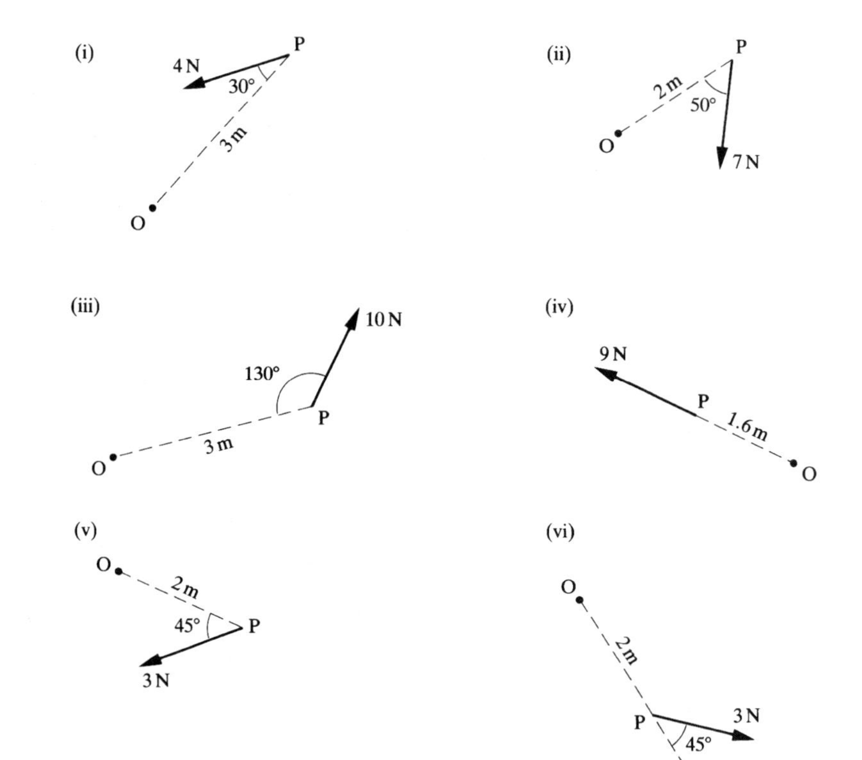

2. In the diagram, two people are moving a bed. Explain, using moments, how you can tell which one is pulling and which one is pushing.

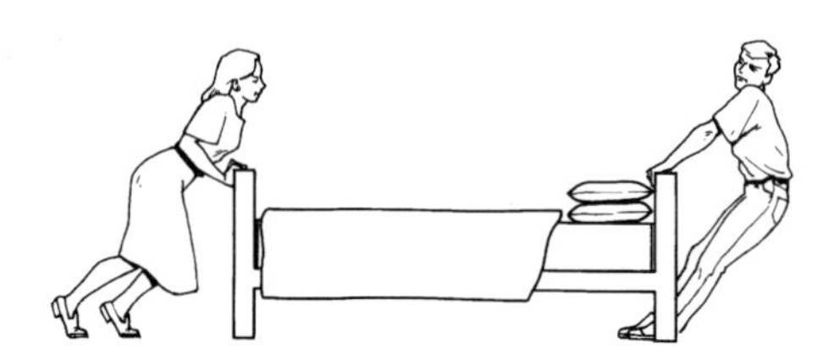

3. The diagram shows three children pushing a playground roundabout. Hannah and David want it to go one way but Rabina wants it to go the other way. Who win(s)?

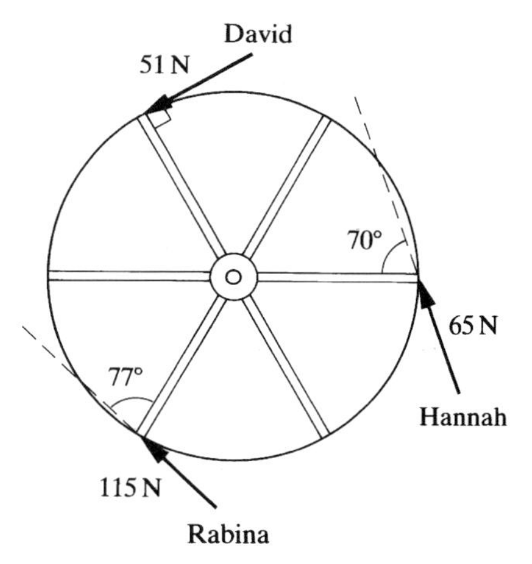

4. The operating instructions for a small crane specify that when the jib is at an angle of $25°$ above the horizontal, the maximum safe load for the crane is $5000\ \text{kg}$. Assuming that this maximum load is determined by the maximum moment that the pivot can support, what is the maximum safe load when the angle between the jib and the horizontal is:

(i) 40°, (ii) 60°, (iii) an angle θ?

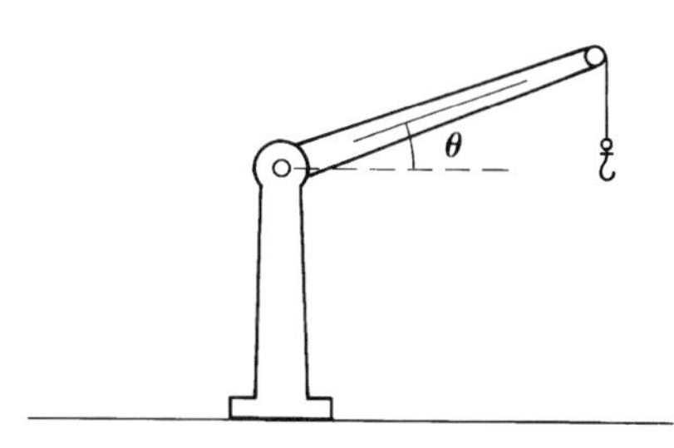

5. In each of these diagrams, a uniform beam of mass $5\ \text{kg}$ and length $4\ \text{m}$, freely hinged at one end, A, is in equilibrium. Find the magnitude of the force T in each case.

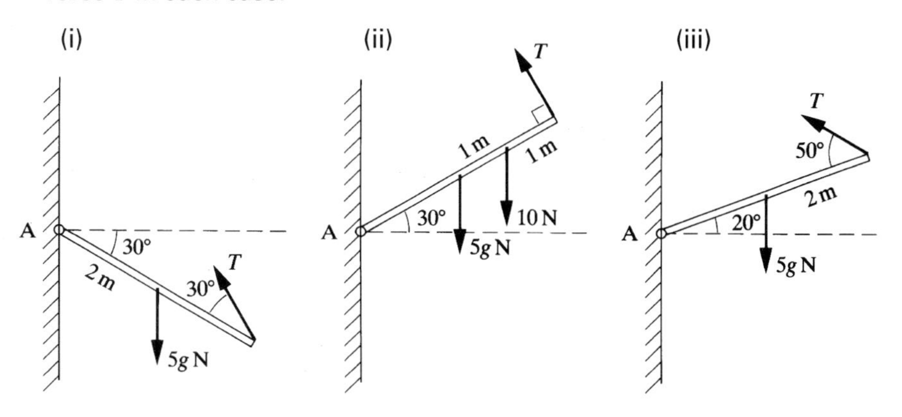

Exercise 2B continued

6. The diagram shows a uniform rectangular sign ABCD, 3 m × 2 m, of weight 20 N. It is freely hinged at A and supported by the string CE, which makes an angle of 30° with the horizontal. The tension in the string is T in Newtons.

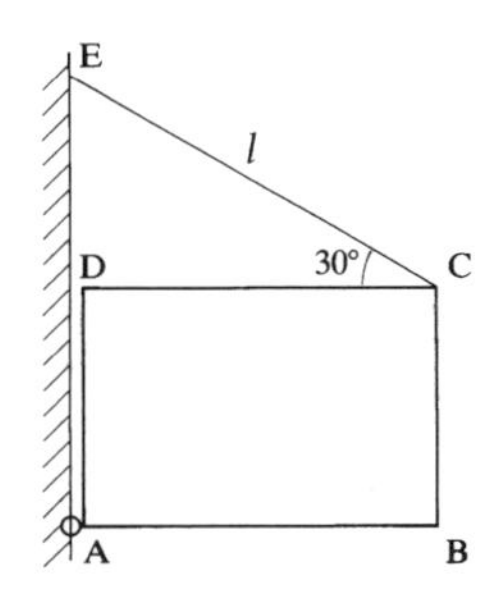

(i) Resolve the tension T into horizontal and vertical components.

(ii) Hence show that the moment of the tension in the string about A is given by

$$2T\cos 30° + 3T\sin 30°$$

(iii) Write down the moment of the sign's weight about A.

(iv) Hence show that $T = 9.28$.

(v) Hence find the horizontal and vertical components of the reaction on the sign at the hinge, A.

You can also find the moment of the tension in the string about A as $d \times T$, where d is the length of AF as shown in the diagram.

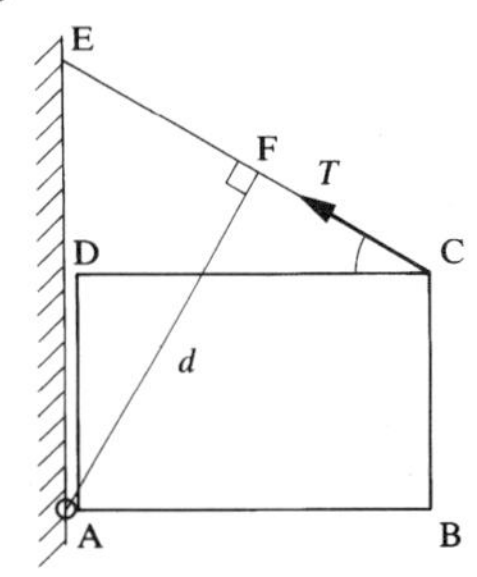

(vi) Find (a) the angle ACD, (b) the length d

(vii) Show that you get the same value for T when it is calculated in this way.

7. The diagram shows a simple crane. The weight of the jib (AB) may be ignored. The crane is in equilibrium in the position shown.

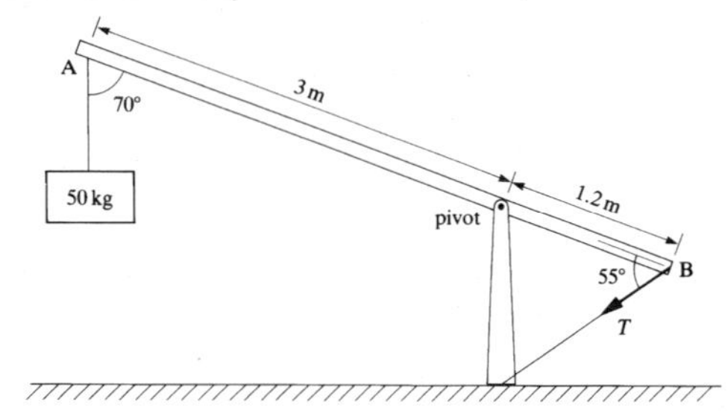

(i) By taking moments about the pivot, find the magnitude of the tension T in newtons.

(ii) Find the reaction of the pivot on the jib in the form of components parallel and perpendicular to the jib.

(iii) Show that the total moment about the end A of the forces acting on the jib is zero.

(iv) What would happen if

(a) the rope holding the $50\ \mathrm{kg}$ mass snapped;

(b) the rope with tension T snapped?

8. The diagram shows the pedals on a bicycle.

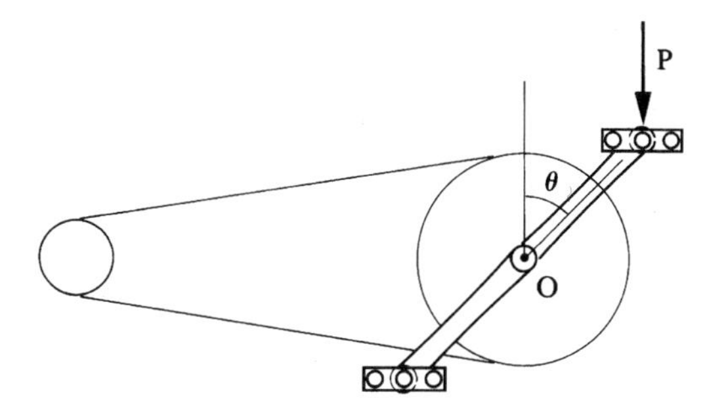

(i) Assume that the force P exerted by the cyclist on one pedal while it is travelling downwards is constant and that it is zero whilst the pedal moves upwards. Draw sketch graphs to show how the moment about the axle varies as the angle θ varies over several turns:

(a) if the force from only one leg is considered;

(b) if the forces from both legs are considered.

(ii) Does your experience of cycling lead you to think that the assumption of a constant downwards force on the pedal is reasonable?

(iii) What is the effect of having toe-clips on the pedals?

9. A uniform plank, AB, of mass $50\ \mathrm{kg}$ and length $6\ \mathrm{m}$ is in equilibrium leaning against a smooth wall at an angle of $60°$ to the horizontal. The lower end, A, is on rough horizontal ground.

(i) Draw a diagram showing all the forces acting on the plank.

(ii) Write down the total moment about A of all the forces acting on the plank.

(iii) Find the normal reaction of the wall on the plank at point B.

(iv) Find the frictional force on the foot of the plank. What can you deduce about the coefficient of friction between the ground and the plank?

(v) Show that the total moment about B of all the forces acting on the plank is zero.

Exercise 2B continued

10. A uniform ladder of mass 20 kg and length $2l$ rests in equilibrium with its upper end against a smooth vertical wall and its lower end on a rough horizontal floor. The coefficient of friction between the ladder and the floor is μ. The normal reaction at the wall is S, the frictional force at the ground is F and the normal reaction at the ground is R. The ladder makes an angle α with the horizontal.

For each of the cases, (a) $\alpha = 60°$, (b) $\alpha = 45°$, (c) $\alpha = 75°$:
(i) Draw a diagram showing the forces acting on the ladder.
(ii) Find the magnitudes of S, F and R.
(iii) Find the least possible value of μ.

11. The diagram shows a car's hand brake. The force F is exerted by the hand in operating the brake, and this creates a tension T in the brake cable. The hand brake is freely pivoted at point B.
(i) Draw a diagram showing all the forces acting on the hand brake.
(ii) What is the required magnitude of force F if the tension in the brake cable is to be 1000 N?
(iii) A child applies the hand brake with a force of 10 N. What is the tension in the brake cable?

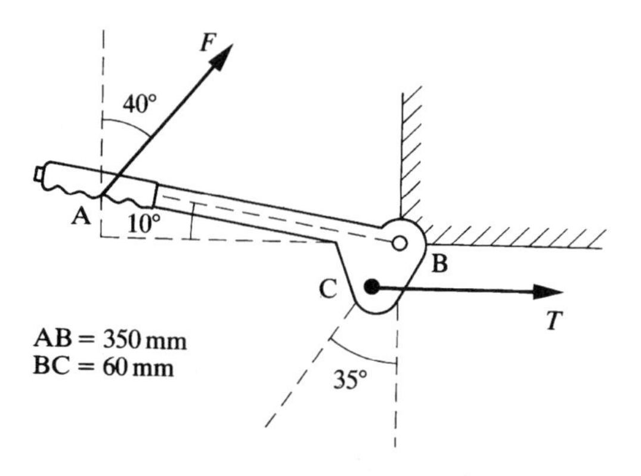

12. The diagram shows four tugs manoeuvring a ship. A and C are pushing it, B and D are pulling it.

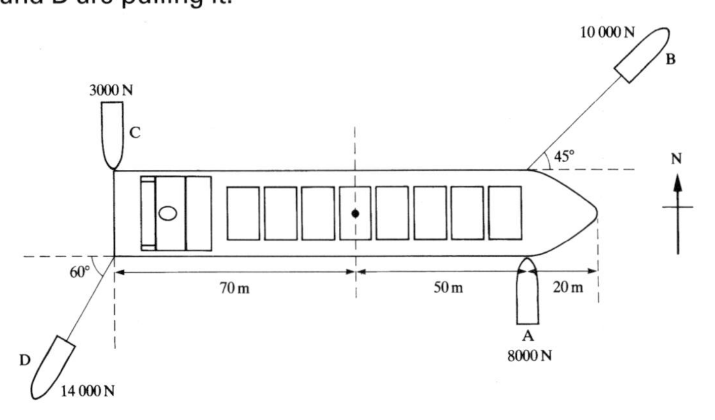

(i) Show that the resultant force on the ship is less than 100 N.

(ii) Find the overall turning moment on the ship about its centre point, O.

A breeze starts to blow from the South, causing a total force of 2000 N to act uniformly along the length of the ship, at right angles to it.

(iii) Is it possible (assuming B and D continue to apply the same forces) for tugs A and C to counteract the sideways force on the ship by altering the forces with which they are pushing, while maintaining the same overall moment about the centre of the ship?

13. The diagram shows a uniform girder AB of weight 3000 N and lengh 6 m which has been hoisted into the air by a crane. The lengths of the ropes AC and BC are both 5 m. The tension in AC is T_1N, that in BC T_2N. The girder makes an angle α with the horizontal. The point X is directly below C and M is the mid-point of AB.

Fred, whose weight is 1000 N, was sitting on the girder when it was hoisted and now finds himself in mid-air. At the time of the question the girder and Fred are stationary. Fred is at point F where AF = 4 m and BF = 2 m.

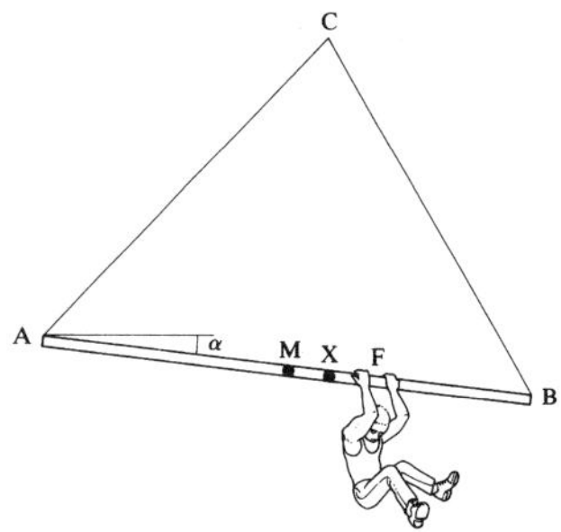

(i) Draw a diagram showing the forces acting on the girder.

(ii) By taking moments about C find the distances MX and FX.

(iii) What is the significance of the point X?

(iv) Find the values of α, T_1 and T_2.

(v) Fred decides to try to reach the point B. What is the value of α when he gets there?

14. The diagram shows a uniform ladder AB of mass m and length $2l$ resting in equilibrium with its upper end A against a smooth vertical wall and its lower end B on a smooth inclined plane. The inclined plane makes an angle θ with the horizontal and the ladder makes an angle ϕ with the wall.

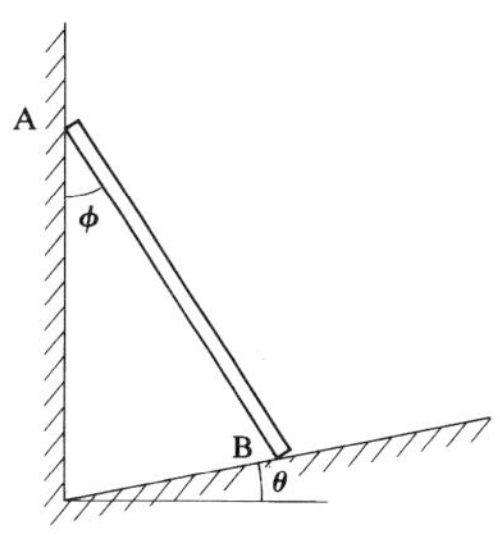

Exercise 2B continued

(i) Find the value of ϕ when θ equals

(a) $10°$ (b) $20°$ (c) $30°$

(ii) What is the relationship between ϕ and θ?

Sliding and toppling

For Discussion

The photograph shows a double decker bus on a test ramp. The angle of the ramp to the horizontal is slowly increased.

What happens to the bus? Would a loaded bus behave differently from the empty bus in the photograph?

Experiment

The diagrams show a force being applied to a cereal packet in different positions.

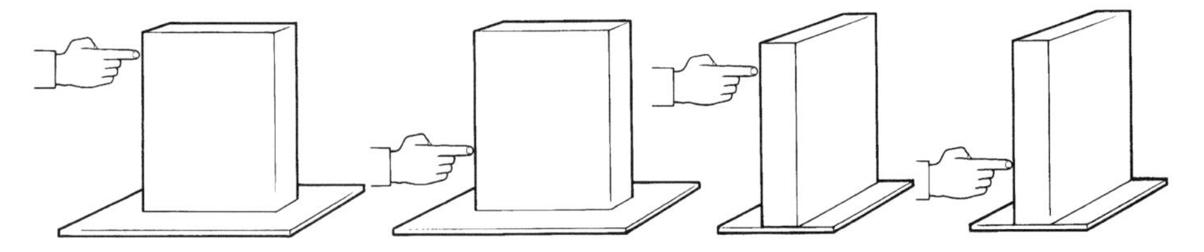

In which case do you think the packet is most likely to fall over? In which case is it most likely to slide? Investigate your answers practically, using boxes of different shapes.

The diagram overleaf show the cereal packet placed on a slope. As the angle of the slope to the horizontal increases the box will either topple or slide. Which is more likely to slide? Which is more likely to topple?

Experiment continued

Investigate your answers using boxes of different shapes. To what extent is this situation comparable to that of the bus on the test ramp?

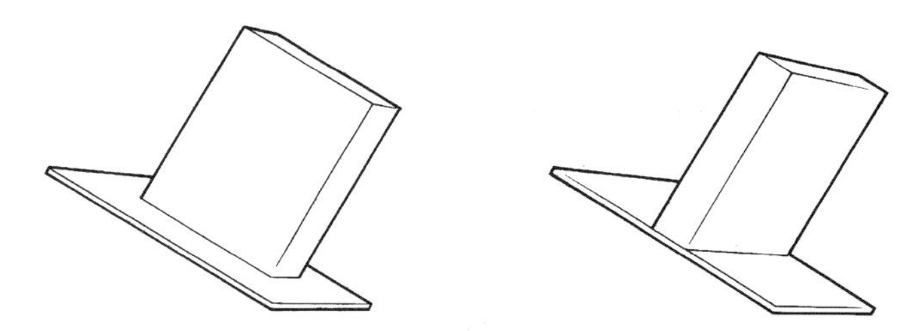

Two critical cases

The results of experiments like these can be predicted by considering the forces and moments that are acting in these two situations.

(i) *The object is on the point of sliding.*

In this situation, according to our model for friction, $F = \mu R$

(ii) *The object is on the point of toppling*

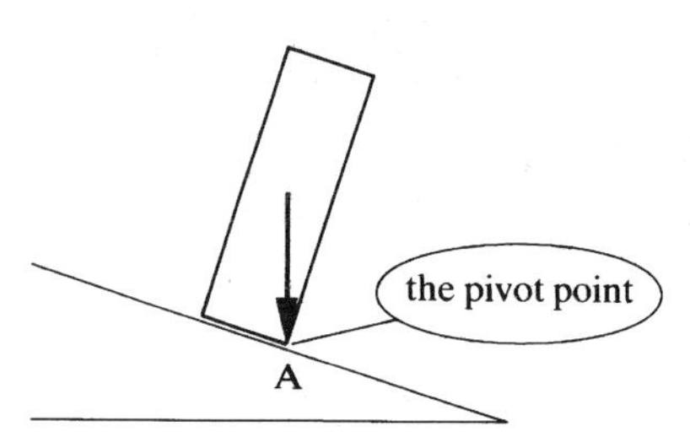

Figure 2.16

In this situation the pivot is the lowest point A. In this critical case:

- the centre of mass is directly above A so that the weight acts vertically downwards through A (figure 2.16);
- the total reaction of the plane on the object acts through A, vertically upwards; this is the resultant of the normal reaction of the plane on the object and the frictional force.

You may find it helpful to visualise A as the only point of contact between the plane and the object. Visually you think of the reaction force acting within the area of contact but it acts at the edge when toppling is about to occur.

Once the object is actually toppling, of course, there really is only one point of contact (figure 2.17).

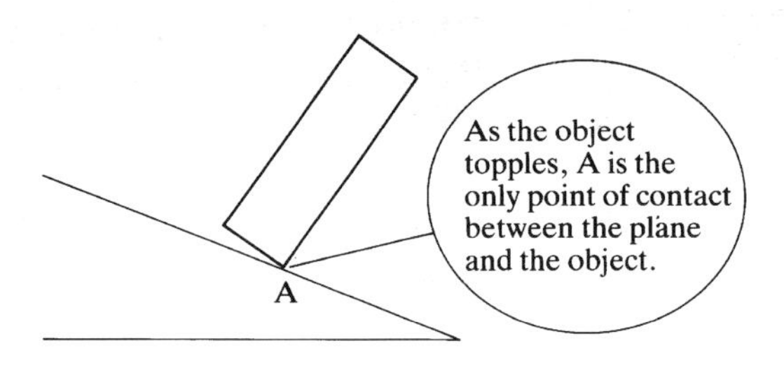

Figure 2.17

EXAMPLE

An increasing force P newtons is applied to a block, as shown in the diagram, until the block moves. The coefficient of friction between the block and the plane is 0.4. Does it slide or topple?

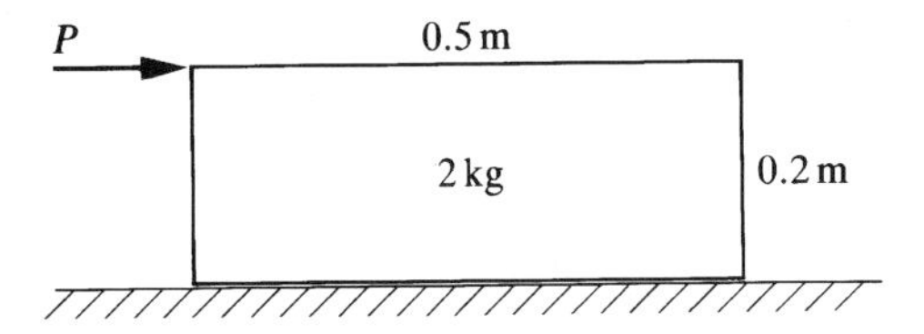

Solution

The forces acting are shown on the diagram. The normal reaction may be thought of as a single force acting somewhere within the area of contact. When toppling occurs (or is about to occur) the line of action is through the edge about which it topples.

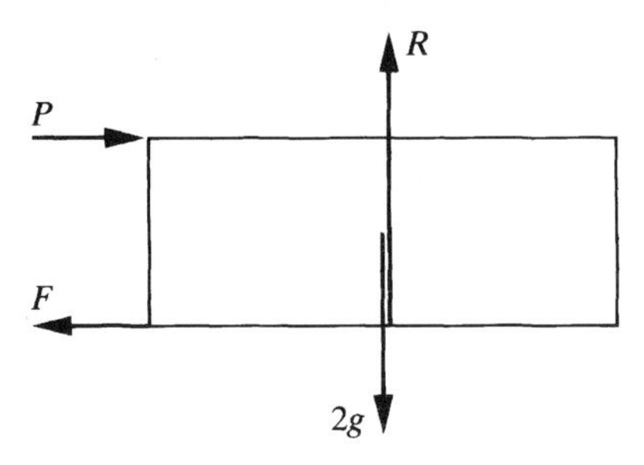

Until the block moves, it is in equilibrium.

Horizontally, $P = F$

Vertically, $R = 2g$

If *sliding* is about to occur, $F = \mu R$

$$P = \mu R = 0.4 \times 2g$$
$$= 7.84$$

If the block is about to *topple*, then A is the pivot point and the reaction of the plane on the block acts at A. Taking moments about A gives

$$2g \times 0.25 - P \times 0.2 = 0$$
$$P = 24.5$$

So to slide P needs to exceed 7.84 N but to topple it needs to exceed 24.5 N: the block will slide before it topples.

EXAMPLE

A block of mass 3 kg is placed on a slope as shown below. The angle α is gradually increased. What happens to the block, given that the coefficient of friction between the block and slope is 0.6?

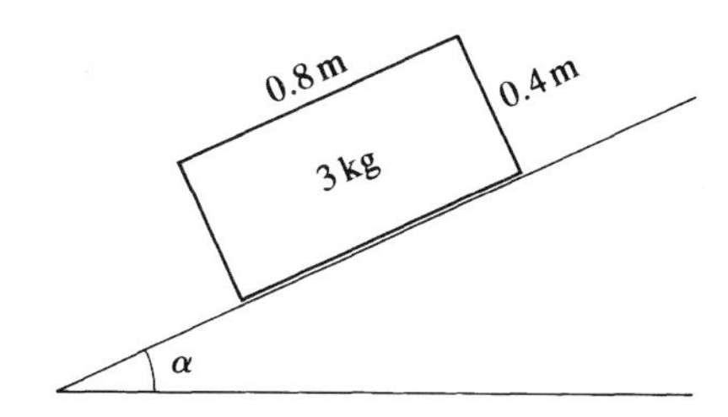

Solution

The diagram shows the forces acting.

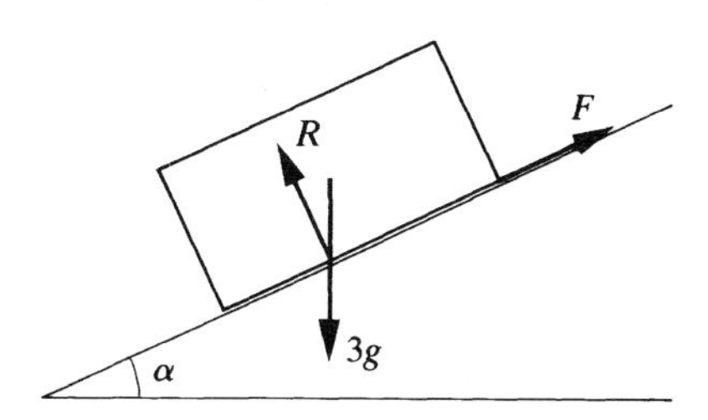

Resolving the weight parallel and perpendicular to the slope, and applying equilibrium equations:

Parallel to the slope: $F = 3g\sin\alpha$

Perpendicular to the slope: $R = 3g\cos\alpha$

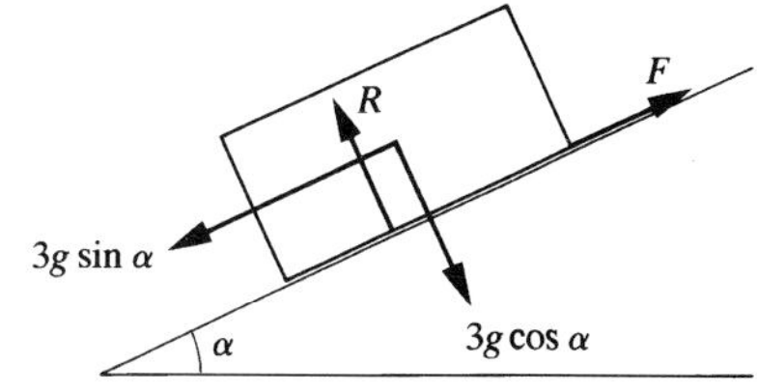

Dividing gives

$$\frac{F}{R} = \tan\alpha \Rightarrow F = R\tan\alpha$$

When the block is on the point of slipping, $F = \mu R$, so

$$R\tan\alpha = \mu R$$
$$\tan\alpha = \mu = 0.6 \ \therefore \alpha = 31°$$

The block is on the point of slipping when $\alpha = 31°$.

When the block is on the point of toppling about the corner A, the centre of mass, G, is directly above A, as shown below.

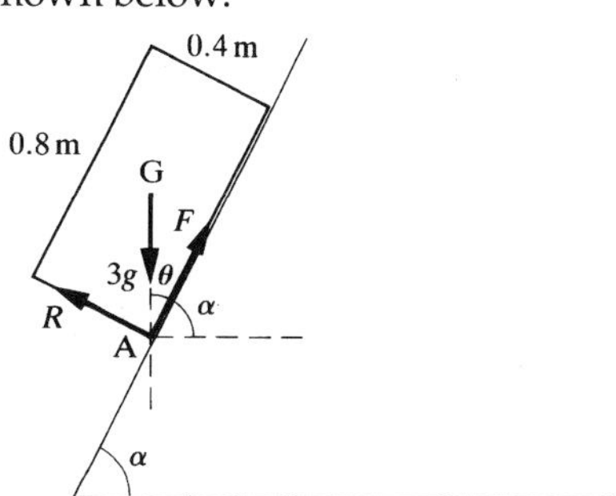

The angle θ is given by

$$\tan\theta = \frac{0.4}{0.8}$$
$$\theta = 26.6°.$$

Now $\alpha = 90° - \theta = 63.4°$, so the block will topple when $\alpha = 63.4°$.

The angle for sliding, 31° is less than the angle for toppling, 63.4°, so the block slides when $\alpha > 31°$.

Exercise 2C

1. A force of magnitude P newtons acts as shown on a block resting on a horizontal plane. The coefficient of friction between the block and the plane is 0.7.

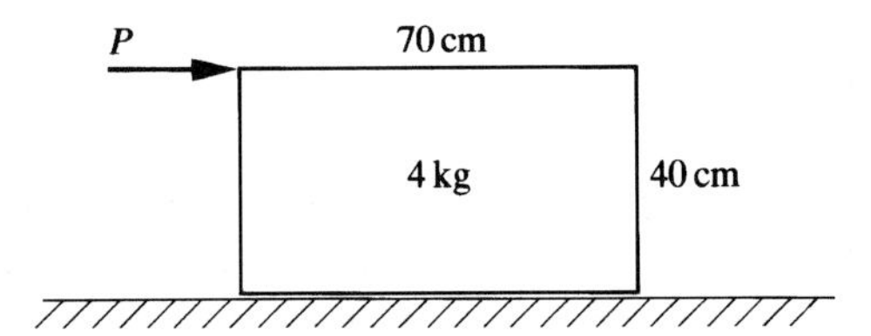

The magnitude of the force P is gradually increased from zero.

(i) Find the magnitude of P if the block is on the point of sliding assuming it does not topple.

(ii) Find the magnitude of P if the block is on the point of toppling assuming it does not slide.

(iii) Does the block slide or topple?

2. A solid uniform cuboid is placed on a horizontal surface. A force P is applied as shown in the diagram.

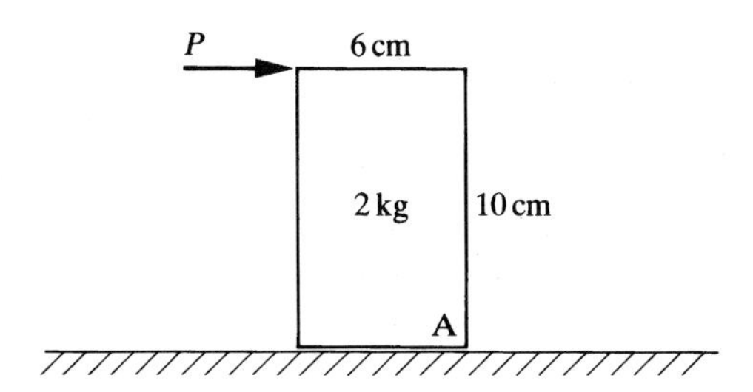

(i) If the block is on the point of sliding express P in terms of μ, the coefficient of friction between the block and the plane.

(ii) Find the magnitude of P if the cuboid is on the point of toppling.

(iii) For what values of μ will the block slide before it topples?

(iv) For what values of μ will the block topple before it slides?

3. A horizontal force of increasing magnitude is applied to the middle of the face of a 50 cm uniform cube, at right angles to the face. The coefficient of friction between the cube and the surface is 0.4 and the cube is on a level surface. What happens to the cube?

4. A uniform cube, of side 50 cm and weight 1200 N, is situated on a rough horizontal plane. The coefficient of friction between the cube and the plane is 0.4. A horizontal force is applied at the middle of one of the top edges of the cube, at right angles to the edge. The force is increased until the cube either slides or topples.

(i) Assuming that it does not topple, what force is needed to make the cube slide?

(ii) Assuming that it does not slide, what force is needed to make the cube topple?

(iii) Which happens first, sliding or toppling?

5. A cube of side 4 cm and weight 60 N is situated on a rough horizontal plane. The coefficient of friction between the cube and the plane is 0.4. A force P N acts in the middle of one of the edges of the top of the cube, at right angles to it and at an angle θ to the upward vertical.

In the cases when the value θ is (a) $60°$, (b) $80°$, find

(i) the force P needed to make the cube slide, assuming it does not topple;

(ii) the force P needed to make the cube topple, assuming it does not slide;

(iii) whether it first slides or topples as the force P is increased.

For what value of θ do toppling and sliding occur for the same value of P, and what is that value of P?

6. A uniform rectangular block of height 30 cm and width 10 cm is placed on a rough plane inclined at an angle α to the horizontal. The box lies on the plane with its length horizontal. The coefficient of friction between the box and the plane is 0.25.

(i) Assuming that it does not topple, for what value of α does the block just slide?

(ii) Assuming that it does not slide, for what value of α does the block just topple?

(iii) The angle α is increased slowly from an initial value of $0°$. Which happens first, sliding or toppling?

7. A solid uniform cuboid, 10 cm $\times$ 20cm $\times$ 50cm, is to stand on an inclined plane, which makes an angle α with th horizontal. One edge of the

Exercise 2C continued

cuboid is to be parallel to the line of the slope. The coefficient of friction between the cuboid and the plane is μ.

(i) Which face of the cuboid should be placed on the slope to make it (a) least likely and (b) most likely to topple?

(ii) How does the cuboid's orientation influence the likelihood of it sliding?

(iii) Find the range of possible values of μ in terms of α in the situations where:

(a) it will slide whatever its orientation;

(b) it will topple whatever its orientation.

8. A cube of side 4 cm and mass 100 g is acted on by a force as shown in the diagram.

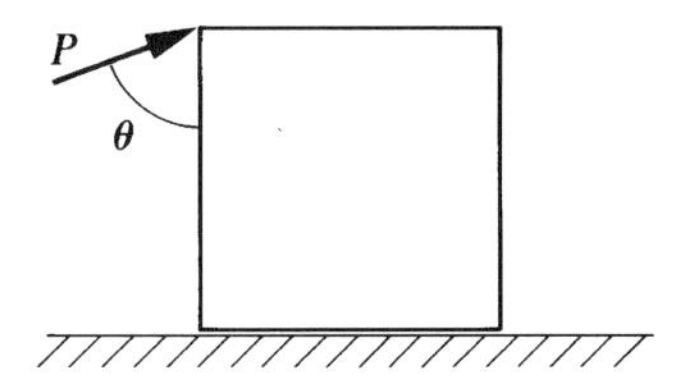

The coefficient of friction between the cube and the plane is 0.5. What happens to the cube if

(i) $\theta = 45°$ and $P = 1\text{ N}$,

(ii) $\theta = 60°$ and $P = 0.5\text{ N}$?

Investigations

Baby buggy

Borrow a baby buggy and investigate its stability. How stable is it when you hang some shopping on its handle?

How could the design of the buggy be altered to improve its stability?

Think about the handling of the buggy in other situations. Would your changes cause any problems?

Whitewashing a house

WARNING: When you carry out experiments for this investigation, do them using a home-made physical model. It is dangerous to carry them out for real.

A man is whitewashing his house but finds that his ladder is not quite long enough for him to reach the top of a gable. He places an old table near the wall of the house and rests the foot of the ladder on it, having first nailed a block of wood onto it to stop the ladder slipping.

What happens when he climbs the ladder?

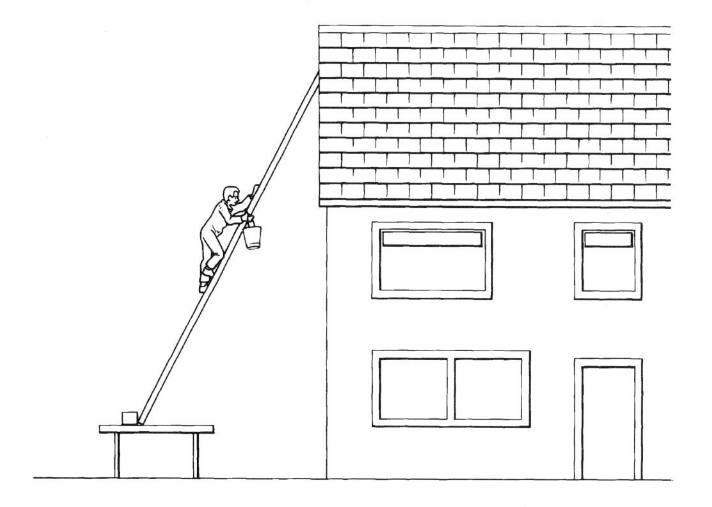

Toolbox

Which tools in a typical tool box depend upon moments for their successful operation?

The Dartford Bridge

The Dartford Bridge carries five lanes of M25 traffic across the River Thames at Dartford. The diagram below shows a drawing of one quarter of the bridge based on the assumption that the deck is horizontal. The mass of each deck section is 1.3×10^6 kg. Investigate the forces acting on each component of this bridge, i.e. the towers, the cables and the deck sections. Assume that a cable is attached to the centre of each deck section.

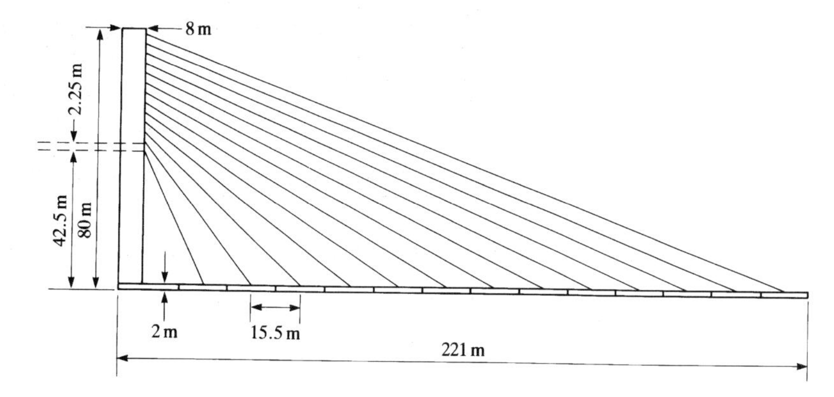

Sliding and Toppling

Make a pile of rough bricks on a board, then raise one edge of the board so that it slopes. Investigate what happens as the angle of the slope is increased.

KEY POINTS

- The moment of a force F about a point O is given by the product Fd where d is the perpendicular distance from O to the line of action of the force.

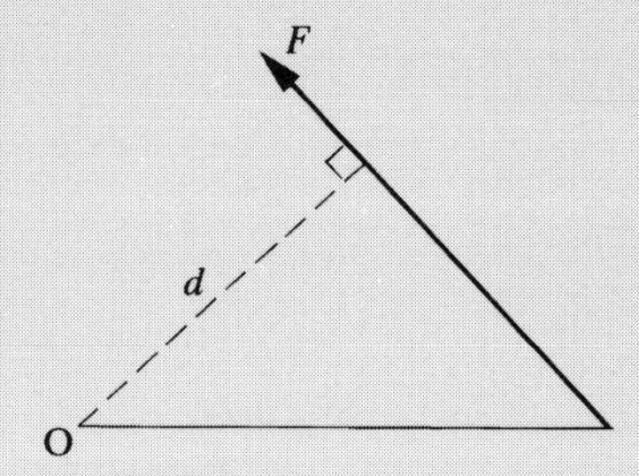

- Anticlockwise moments are usually called positive, clockwise negative.

- If a body is in equilibrium the sum of the moments about any point of the forces acting on it is zero.

3 Centre of mass

Let man then contemplate the whole of nature in her full and grand mystery . . . It is an infinite sphere, the centre of which is everywhere, the circumference nowhere.

Blaise Pascal

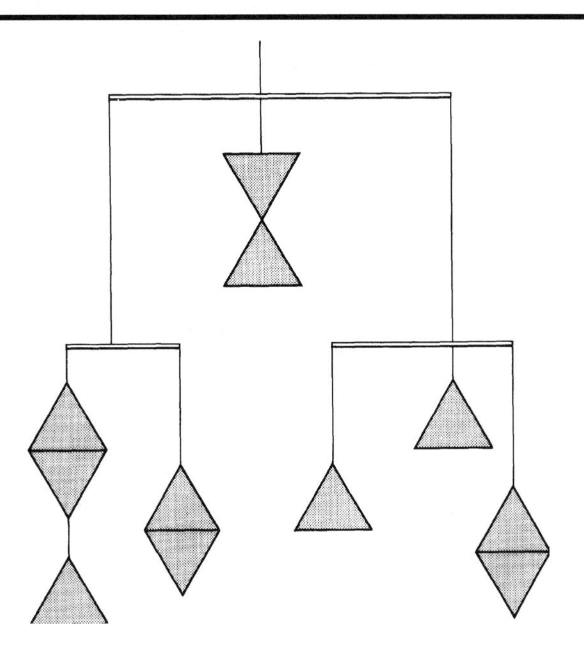

The diagram, which is drawn to scale, shows a mobile suspended from the point P. The horizontal rods and the strings are light but the geometrically shaped pieces are made of uniform heavy card. Does the mobile balance? If it does, what can you say about the position of its centre of mass?

How does the position of the centre of mass of the mobile compare with that of the gymnast in the picture (right)?

You have met the concept of centre of mass in the context of two general models.

- *The particle model*

  The centre of mass is the single point at which the whole mass of the body may be taken to be situated.

- *The rigid body model*

  The centre of mass is the balance point of a body with size and shape.

The following examples show how to calculate the position of the centre of mass of a body.

EXAMPLE

An object consists of three point masses 8 kg, 5 kg and 4 kg attached to a rigid light rod as shown.

8 kg   1.2 m   5 kg   0.6 m   4 kg
O

Calculate the distance of the centre of mass of the object from end O. (Ignore the mass of the rod).

Solution

Suppose the centre of mass C is $\bar{x}$ m from O: if a pivot were at this position the rod would balance.

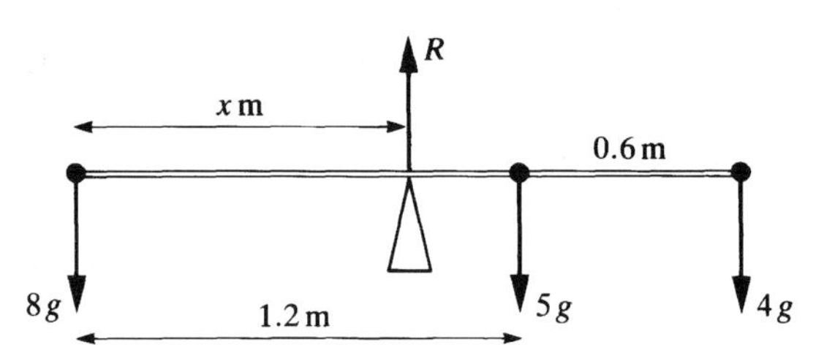

For equilibrium $R = 8g + 5g + 4g = 17g$ newtons.

Taking moments of the forces about O gives:

Total clockwise moment $= (8g \times 0) + (5g \times 1.2) + (4g \times 1.8)$
$= 13.2g$ Nm

Total anticlockwise moment $= Rx$
$= 17g\bar{x}$ Nm.

The overall moment must be zero for the rod to be in balance, so

$$17g\bar{x} - 13.2g = 0$$

$$\Rightarrow \qquad \bar{x} = \frac{13.2}{17} = 0.776.$$

The centre of mass is 0.776 m from the end O of the rod.

Note that although the force of gravity was included in the calculation, it cancelled out. The answer depends only on the masses and their distances from the origin and not on the value of g. This leads to the following definition for the position of the centre of mass.

Definition

Consider a set of n point masses $m_1, m_2, \ldots m_n$ attached to a rigid light rod (whose mass is neglected) at positions $x_1, x_2, \ldots x_n$ from one end O. The situation is shown in figure 3.1.

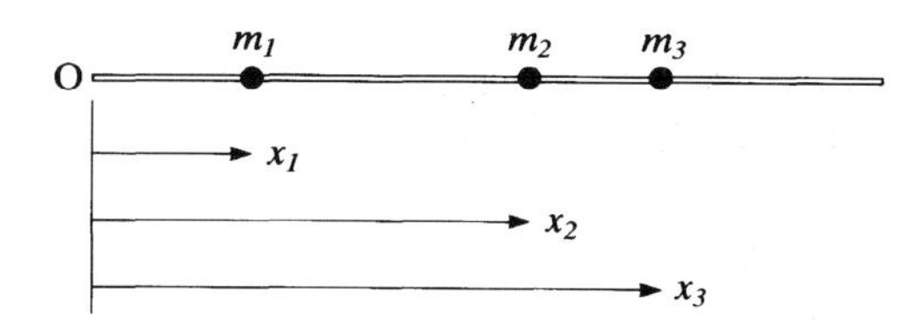

Figure 3.1

The position of the centre of mass from O, labelled $\bar{x}$, is defined by the equation

moment of whole mass at centre of mass = sum of moments of individual masses

or
$$M\bar{x} = \sum_{c=1}^{n} x_i m_i$$

where M is the total mass (or $\Sigma\, m_i$).

EXAMPLE

A uniform rod of length 2 m has mass 5 kg. Masses of 4 kg and 6 kg are fixed at each end of the rod. Find the centre of mass of the rod.

Solution

Since the rod is uniform, it can be treated as a point mass at its centre. The diagram illustrates this situation.

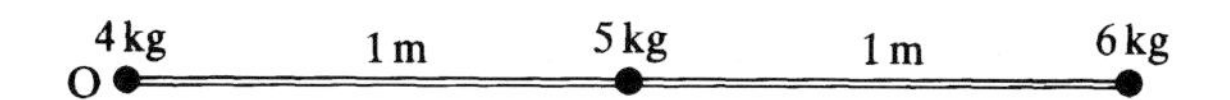

Taking the left hand end of the rod as origin,

$$M\bar{x} = \sum m_i x_i$$
$$(4+5+6)\bar{x} = 4\times 0 + 5\times 1 + 6\times 2$$
$$15\bar{x} = 17$$
$$\bar{x} = \frac{17}{15}$$
$$= 1\frac{2}{15}$$

So the centre of mass is 1.133 m from the 4 kg point mass.

NOTE *This definition of centre of mass, $M\bar{x} = \Sigma m_i, x_i$, has moved away from the intuitive idea of the centre of mass as the balance point of the object. You can see that the two are equivalent in the following (rather heavy-handed) alternative solution.*

Let the position of the centre of mass be point G, at distance $\bar{x}$ from O.

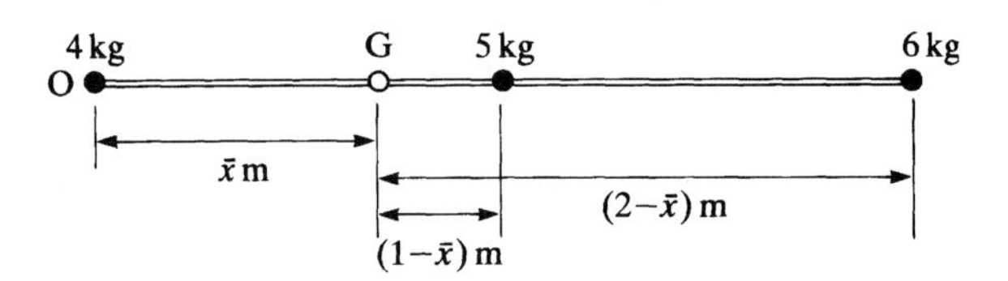

The distance of G from each point mass is as shown in the diagram.

Taking moments about G:

$$4g\times\bar{x} - 5g\times(1-\bar{x}) - 6g\times(2-\bar{x}) = 0$$
$$4\times\bar{x} - 5\times 1 + 5\bar{x} + 5\bar{x} - 6\times 2 + 6\bar{x} = 0$$
$$(4+5+6)\bar{x} = 5\times 1 + 6\times 2$$

which is the equation found by using

$$M\bar{x} = \sum x_i m_i$$

and of course leads to the same answer, $\bar{x} = 1\frac{2}{15}$.

Composite bodies

The position of the centre of mass of a composite body such as a cricket bat, tennis racquet or golf club is important to sports people who like to feel its balance. If the body is symmetric then the centre of mass will lie on the axis of symmetry. The next example shows how to model a composite body as a system of point masses so that the methods of the previous section can be used to find the centre of mass.

EXAMPLE

A squash racquet of mass 200 g and total length 70 cm consists of a handle of mass 150 g whose centre of mass is 20 cm from the end, and a frame of mass 50 g whose centre of mass is 55 cm from the end.

Find the distance of the centre of mass from the end of the handle.

Solution

The diagram shows the squash racquet and its dimensions.

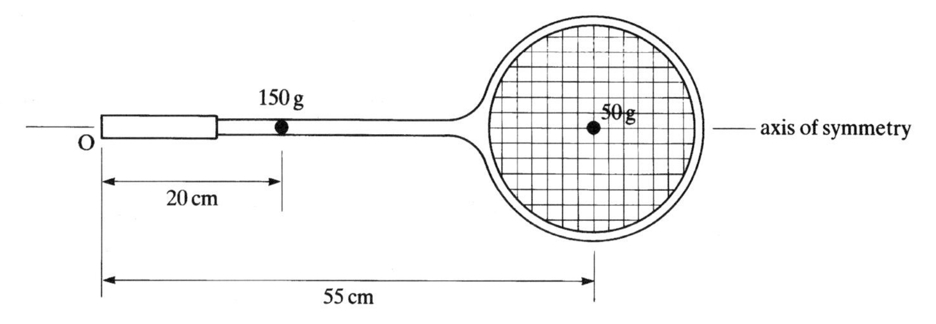

The centre of mass lies on the axis of symmetry, a line running along the middle of the handle and through the centre of the head. We model the frame as a point mass of 0.05 kg a distance 0.55 m from the end O, and the handle as a point mass of 0.15 kg a distance 0.2 m from O.

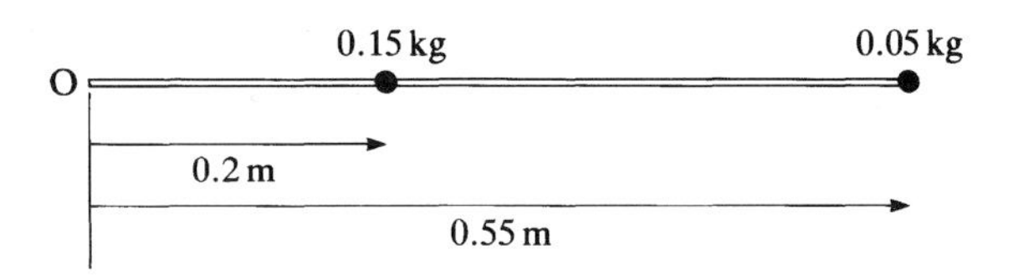

The distance, x_G of the centre of mass from O is given by

$$(0.15 + 0.05)\,x_G = (0.15 \times 0.2) + (0.05 \times 0.55)$$

$$x_G = 0.2875.$$

The centre of mass of the squash racquet is 28.75 cm from the end of the handle.

Centres of mass for different shapes

If an object has an axis of symmetry, like the squash racquet in the example above, then the centre of mass lies on it.

The table below gives the position of the centre of mass of other such objects that you may encounter, or wish to include within models of composite bodies.

Body	Position of centre of mass	Diagram
Solid cone or pyramid	$\frac{1}{4}h$ from base	
Hollow cone or pyramid	$\frac{1}{3}h$ from base	
Solid hemisphere	$\frac{3}{8}r$ from base	
Hollow hemisphere	$\frac{1}{2}r$ from base	
Semi-circular laminar	$\frac{4r}{3\pi}$ from base	

Exercise 3A

1. The diagrams show point masses attached to rigid light rods. In each case calculate the position of the centre of mass relative to the point O.

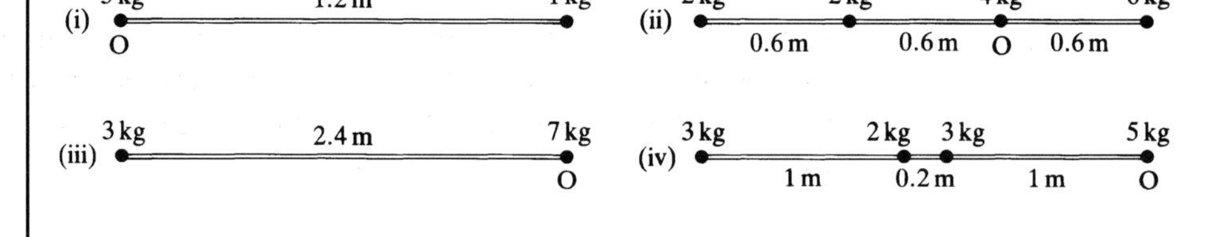

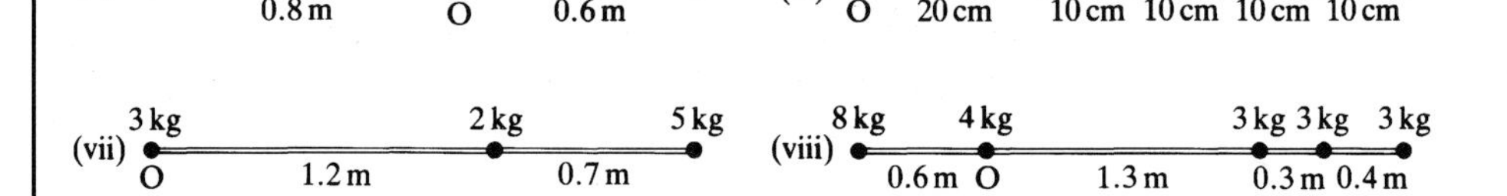

2. A seesaw consists of a uniform plank 4 m long of mass 10 kg. Calculate the centre of mass when two children, of masses 20 kg and 25 kg, sit, one on each end.

3. A weightlifter's bar in a competition has mass 10 kg and length 1 m. By mistake, 50 kg is placed on one end and 60 kg on the other end. How far is the centre of mass of the bar from the centre of the bar itself?

4. A boathook consists of a uniform wooden pole of mass 8 kg and length 3 m. At one end there is a brass hook of mass 1 kg. Find the position of the centre of mass of the boathook. You may assume that, as a consequence of the way it is attached to the pole, the centre of mass of the hook is exactly at the end of the pole.

5. A lollipop lady carries a sign which consists of a uniform rod of length 1.5 m, and mass 1 kg, on top of which is a circular disc of radius 0.25 m and mass 0.2 kg. Find the distance of the centre of mass from the free end of the stick.

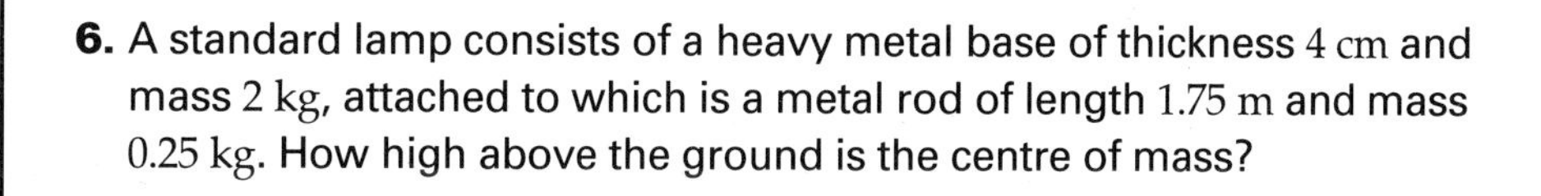

6. A standard lamp consists of a heavy metal base of thickness 4 cm and mass 2 kg, attached to which is a metal rod of length 1.75 m and mass 0.25 kg. How high above the ground is the centre of mass?

7. A rod has length 2 m and mass 3 kg. The centre of mass should be in the middle but due to a fault in the manufacturing process it is not. This error is corrected by placing a 200 g mass 5 cm from the centre of the rod. Where is the centre of mass of the rod itself?

8. The masses of the Earth and the Moon are $5.98 \times 10^{24} \text{ kg}$ and $7.38 \times 10^{22} \text{ kg}$, and the distance between their centres is $3.84 \times 10^{5} \text{ km}$. How far from the centre of the Earth is the centre of mass of the Earth-Moon system?

9. A child's toy consists of four uniform discs, all made out of the same material. They each have thickness 2 cm and their radii are 6 cm, 5 cm, 4 cm and 3 cm. They are placed symmetrically on top of each other to form a tower. How high is the centre of mass of the tower?

Exercise 3A continued

10. A scaffold pole of length 5 m has brackets bolted to it as shown in the diagram below. The mass of the pole is 40 kg and the mass of each bracket is 1 kg. Find the position of its centre of mass.

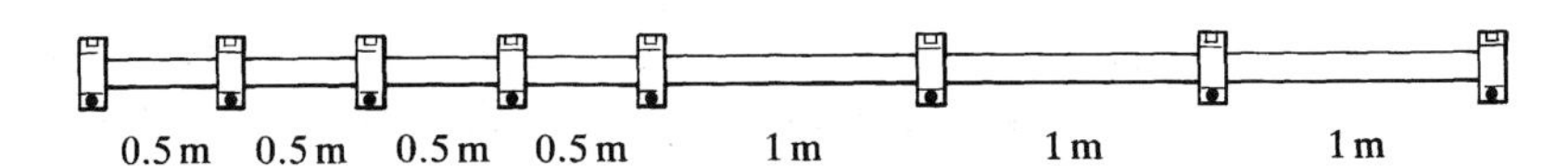

11. An object of mass m_1 is placed at one end of a light rod of length l. An object of mass m_2 is placed at the other end. Find the position of the centre of mass.

12. The diagram illustrates a mobile tower crane. It consists of the main vertical section (mass M tonnes), housing the engine, winding gear and controls, and the boom. The centre of mass of the main section is on its centre line. The boom, which has negligible mass, supports the load (L tonnes) and the counterweight (C tonnes). The main section stands on supports at P and Q, distance $2d$ m apart. The counterweight is held at a fixed distance a m from the centre line of the main section and the load at a variable distance l m.

(i) In the case when $C = 3, M = 10, L = 7, a = 8, d = 2$ and $l = 13$, find the horizontal position of the centre of mass and say what happens to the crane.

(ii) Show that for these values of C, M, a, d and l the crane will not fall over when it has no load, and find the maximum safe load that it can carry.

(iii) Formulate two inequalities in terms of C, M, D, a, d and l that must hold if the crane is to be safe loaded or unloaded.

(iv) Find, in terms of M, a, d and l, the maximum load that the crane can carry.

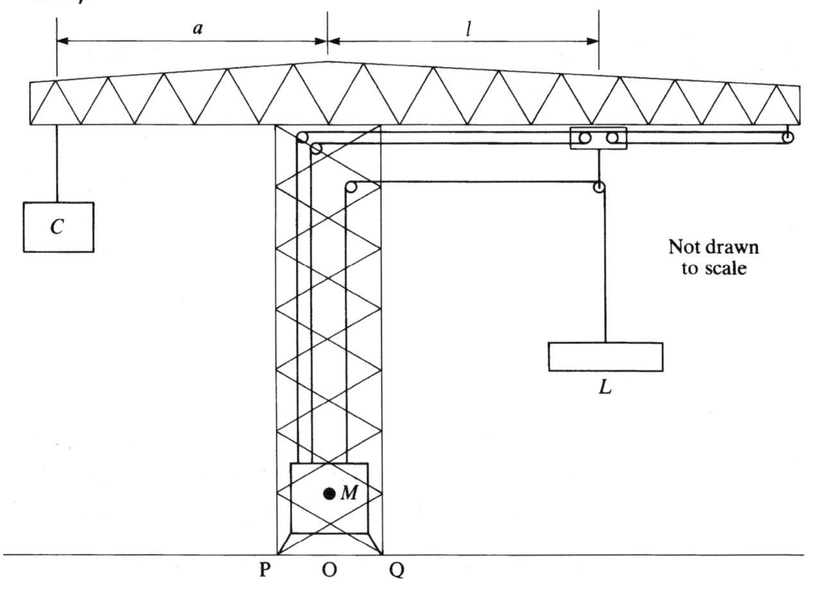

Centre of mass for 2- and 3-dimensional bodies

The techniques developed for finding the centre of mass using moments can be extended into two and three dimensions.

If a two-dimensional body consists of a set of n point masses $m_1, m_2, ..., m_n$ located at positions (x_1, y_1), (x_2, y_2), ..., (x_n, y_n) as in figure 3.2 then the position of the centre of mass of the body $(\bar{x}, \bar{y})$ is given by

$$M\bar{x} = \sum_i m_i x_i \text{ and } M\bar{y} = \sum_i m_i y_i$$

where $M\ (= \sum_i m_i)$ is the total mass of the body.

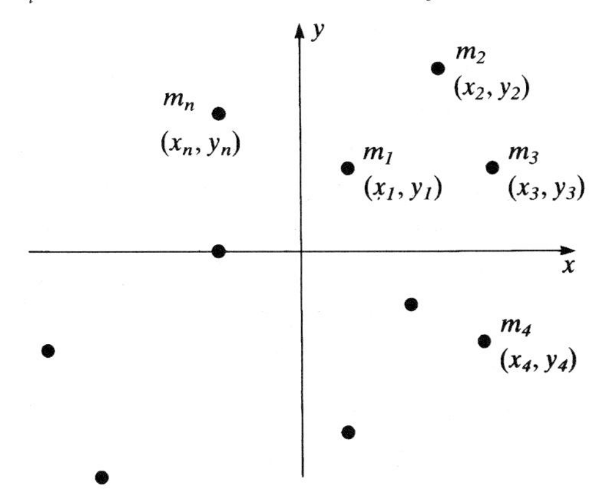

Figure 3.2

In three dimensions, the z-coordinates are also included, to find $\bar{z}$ using

$$M\bar{z} = \sum_i m_i z_i$$

The centre of mass of any composite body in two or three dimensions can be found by replacing each component by a point mass at its centre of mass.

EXAMPLE

Joanna makes herself a pendant in the shape of a letter J made up of rectangular shapes as shown in the diagram. The lengths are shown in cm.

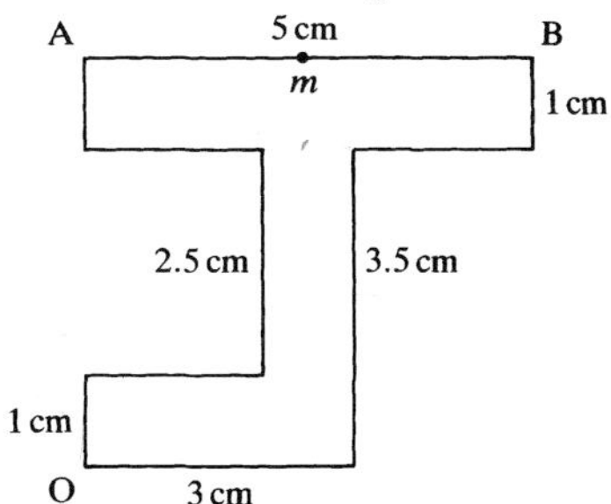

(i) Find the position of the centre of mass of the pendant

(ii) Find the angle that AB makes with the horizontal if she hangs the pendant from a point, M, in the middle of AB.

She wishes to hang the pendant so that AB is horizontal.

(iii) How far along AB should she place the ring that the suspending chain will pass through?

Solution

(i) The first step is to split the pendant into three rectangles. The centre of mass of each of these is at its middle, as shown in the diagram.

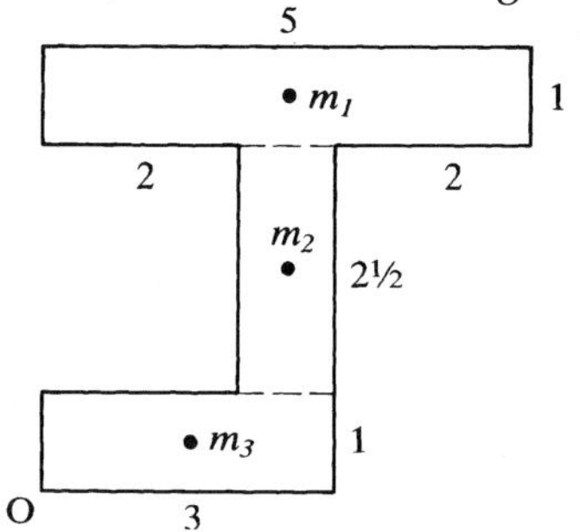

We can model the pendant as three point masses m_1, m_2 and m_3, which are proportional to the areas of the rectangular shapes. Since the areas are 5 cm², 2.5 cm² and 3 cm², the masses, in suitable units, are 5, 2.5 and 3, and the total mass is 5 + 2.5 + 3 = 10.5 (in the same units).

The table below gives the mass and position of m_1, m_2 and m_3.

Mass	Mass units	Position of centre of mass (x, y)
m_1	5	(2.5, 4)
m_2	2.5	(2.5, 2.25)
m_3	3	(1.5, 0.5)
M	10.5	$(\bar{x}, \bar{y})$

Now it is possible to find $\bar{x}$:

$$M\bar{x} = \sum m_i x_i$$
$$10.5\bar{x} = 5 \times 2.5 + 2.5 \times 2.5 + 3 \times 1.5$$
$$\bar{x} = \frac{23.25}{10.5} = 2.2 \text{ cm}$$

Similarly for $\bar{y}$:

$$M\bar{y} = \sum m_i y_i$$
$$10.5\bar{y} = 5 \times 4 + 2.5 \times 2.5 + 3 \times 0.5$$
$$\bar{y} = \frac{27.125}{10.5} = 2.6 \text{ cm}$$

The centre of mass is at (2.2, 2.6)

(ii) When the pendant is suspended, the centre of mass lies immediately below the point of suspension, as shown below.

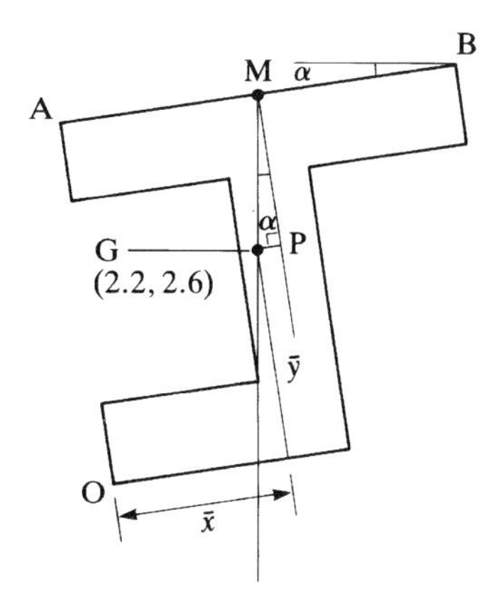

In the diagram G is the centre of mass, and

$$GP = 2.5 - 2.2 = 0.3$$
$$MP = 5 - 2.6 = 2.4$$
$$\therefore \tan\alpha = \frac{0.3}{2.4} \Rightarrow \alpha = 71°$$

AB makes an angle of 7.1° with the horizontal.

(iii) For AB to be horizontal the point of suspension must be directly above the centre of mass, and so it is 2.2 cm from A.

EXAMPLE

Find the centre of mass of a body consisting of a square plate of mass 3 kg and side length 2 m, with small objects of mass 1 kg, 2 kg, 4 kg and 5 kg at the corners of the square.

Solution

The figure shows the square plate, with the origin taken at the corner at which the 1 kg mass is located. The mass of the plate is represented by a 3 kg point mass at its centre.

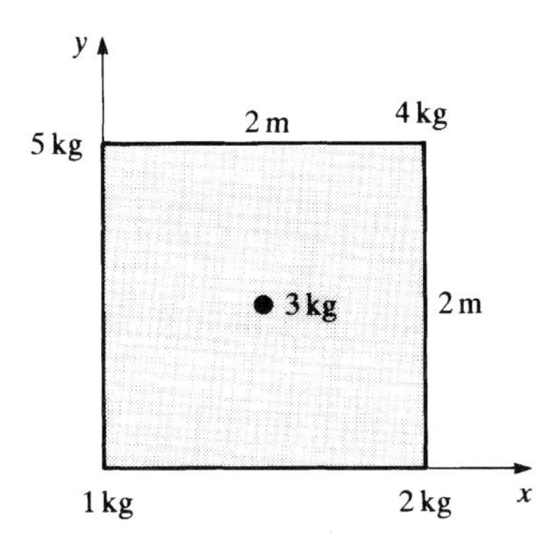

In this example the total mass M (in kilograms) is $1 + 2 + 4 + 5 + 3 = 15$.

The two formulae for $\bar{x}$ and $\bar{y}$ can be combined into one using column vector notation:

$$\begin{bmatrix} M\bar{x} \\ M\bar{y} \end{bmatrix} = \begin{bmatrix} \sum m_i x_i \\ \sum m_i y_i \end{bmatrix}$$

which is equivalent to $$M\begin{pmatrix} \bar{x} \\ \bar{y} \end{pmatrix} = \sum m_i \begin{pmatrix} x_i \\ y_i \end{pmatrix}$$

Substituting our values for M and m_i and x_i and y_i:

$$15\begin{pmatrix} \bar{x} \\ \bar{y} \end{pmatrix} = 1\begin{pmatrix} 0 \\ 0 \end{pmatrix} + 2\begin{pmatrix} 2 \\ 0 \end{pmatrix} + 4\begin{pmatrix} 2 \\ 2 \end{pmatrix} + 5\begin{pmatrix} 0 \\ 2 \end{pmatrix} + 3\begin{pmatrix} 1 \\ 1 \end{pmatrix}$$

$$15\begin{pmatrix} \bar{x} \\ \bar{y} \end{pmatrix} = \begin{pmatrix} 15 \\ 21 \end{pmatrix}$$

$$\begin{pmatrix} \bar{x} \\ \bar{y} \end{pmatrix} = \begin{pmatrix} 1 \\ 1.4 \end{pmatrix}$$

The centre of mass is at the point (1, 1.4).

EXAMPLE

The diagrams show a square table of side 2 m which is supported by 4 legs, each 1 m high, situated the corners of a 1 m square. The mass of each leg is 1 kg and that of the table top 2 kg. The legs and the top are uniform. A small heavy object of mass 6 kg is placed on one corner of the table.

(i) Find the position of the centre of mass relative to the point O shown in the diagram using the axes indicated with unit length 1 m.

(ii) What is the significance of your result?

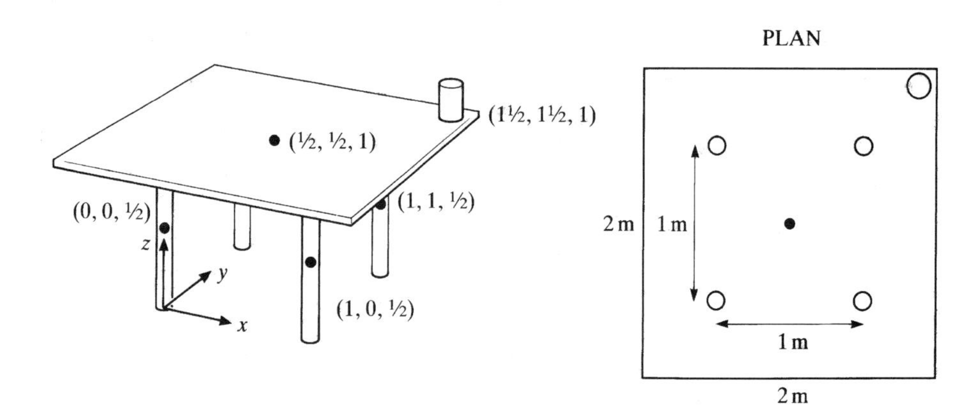

Solution

(i) Let the coordinates of the centre of mass be $(\bar{x}, \bar{y}, \bar{z})$. The total mass is 4×1 (legs) $+ 2$ (top) $+ 6$ (object) $= 12$ kg.

Taking moments about O

$$12 \times \begin{pmatrix} \bar{x} \\ \bar{y} \\ \bar{z} \end{pmatrix} = 1\begin{pmatrix} 0 \\ 0 \\ \frac{1}{2} \end{pmatrix} + 1\begin{pmatrix} 0 \\ 1 \\ \frac{1}{2} \end{pmatrix} + 1\begin{pmatrix} 1 \\ 0 \\ \frac{1}{2} \end{pmatrix} + 1\begin{pmatrix} 1 \\ 1 \\ \frac{1}{2} \end{pmatrix} + 2\begin{pmatrix} \frac{1}{2} \\ \frac{1}{2} \\ 1 \end{pmatrix} + 6\begin{pmatrix} 1\frac{1}{2} \\ 1\frac{1}{2} \\ 1 \end{pmatrix}$$

$$\Rightarrow 12\begin{pmatrix} \bar{x} \\ \bar{y} \\ \bar{z} \end{pmatrix} = \begin{pmatrix} 12 \\ 12 \\ 10 \end{pmatrix}$$

$\Rightarrow$ The centre of mass is at $(1, 1, \frac{5}{6})$

(ii) This means that the centre of mass is in the leg nearest the object and so the table is on the point of toppling over. All the weight is being taken by that one leg.

EXAMPLE

A metal disc of radius 15 cm has a hole of radius 5 cm cut in it as shown below. Find the centre of mass of the disc.

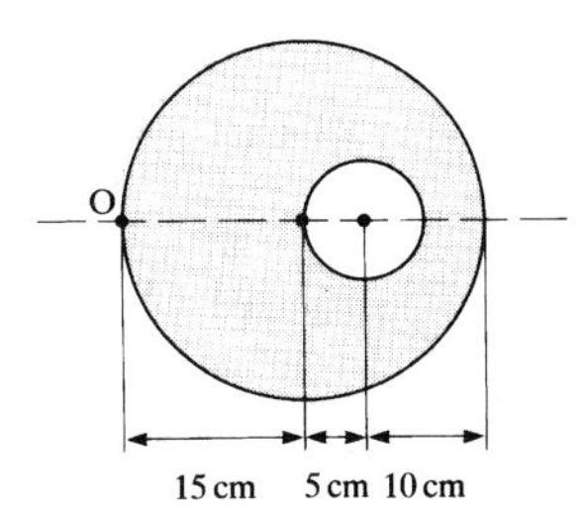

Solution

Think of the original uncut disc as a composite body made up of the final body and a disc to fit into the hole. Since the material is uniform the mass of each part is proportional to its area.

The uncut disc $=$ the final body $+$ the cut out disc.

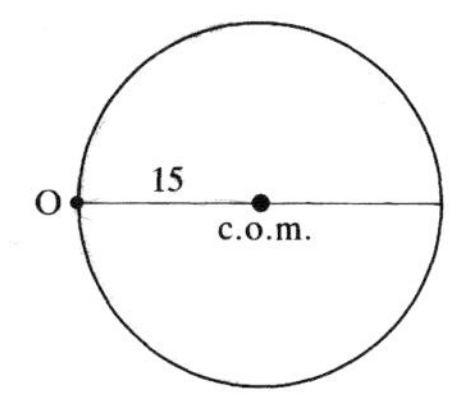

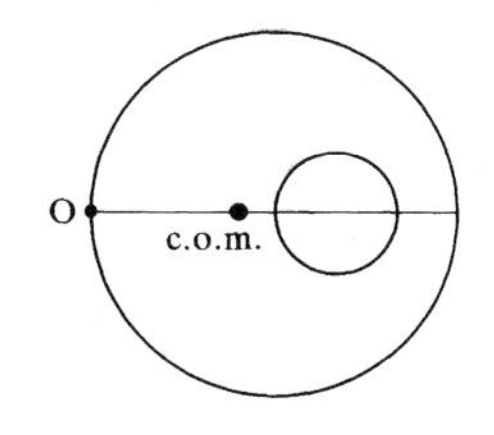

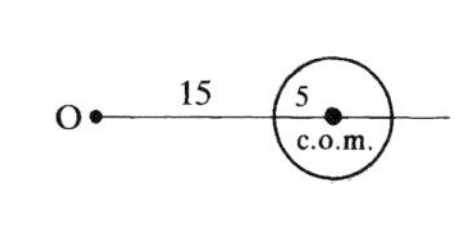

Area	$15^2\pi = 225\pi$	$15^2\pi - 5^2\pi = 200\pi$	$5^2\pi = 25\pi$
Distance from O to centre of mass	15 cm	x cm	20 cm

Taking moments about O

$$225\,\pi \times 15 \quad = \quad 200\,\pi \times x \quad + \quad 25\,\pi \times 20$$

Solving for x gives

$$x = \frac{225 \times 25 \times 20}{200}$$
$$= 14\tfrac{3}{8}$$

The centre of mass is $14\tfrac{3}{8}$ cm from O, that is $\tfrac{5}{8}$ cm to the left of the centre of the disc.

Exercise 3B

1. Find the centre of mass of the following sets of point masses.

(i)

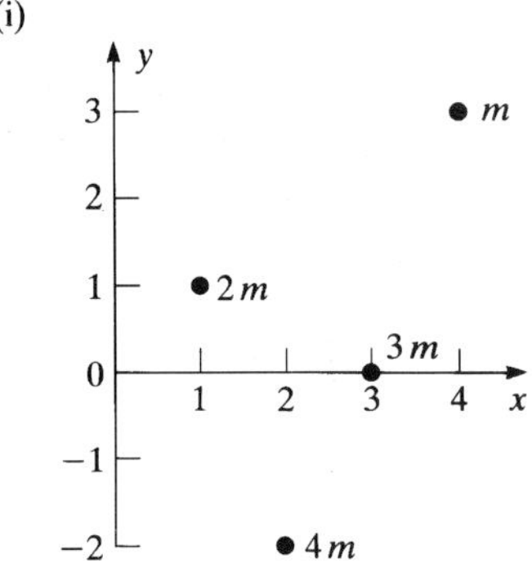

(ii)

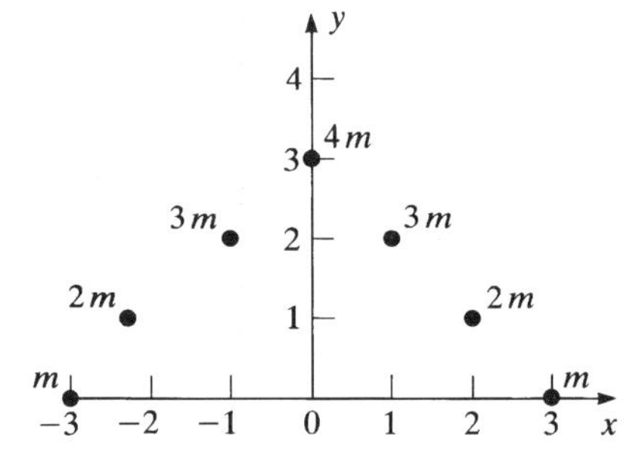

(iii)

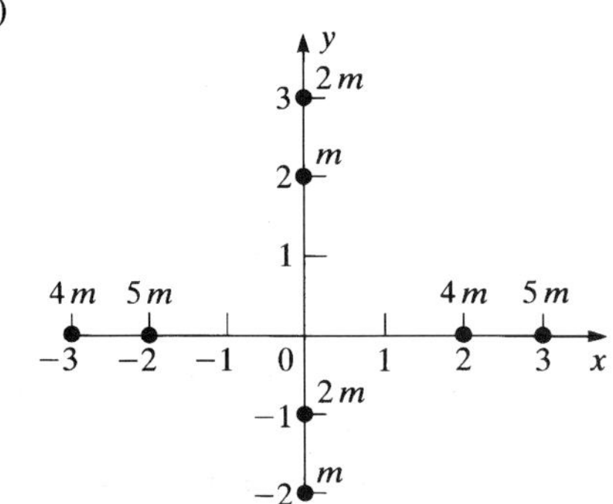

(iv)

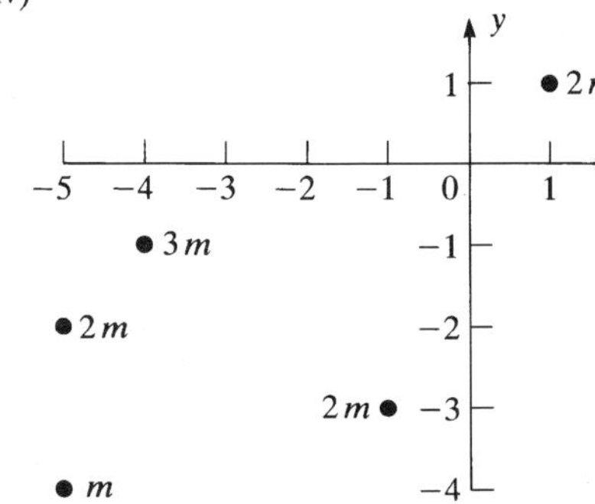

2. Masses of 1, 2, 3 and 4 grams are placed at the corners A, B, C and D of a square piece of uniform cardboard of side 10 cm and mass 5 g. Find the position of the centre of mass relative to axes through AB and AD.

3. The triangle OPQ is made of light material. OP = 40 cm, OQ = 30 cm and angle POQ = 90°. Masses of 0.5 kg, 1.2 kg and 2.3 kg are placed at O, P and Q. Find the position of the centre of mass, relative to axes OP and OQ.

4. As part of a Christmas lights display, letters are produced by mounting bulbs in holders 30 cm apart on light wire frames. The combined mass of a bulb and its holder is 200 g. Find the position of the centre of mass for each of the letters shown below, in terms of its horizontal and vertical displacement from the bottom left hand corner of the letter.

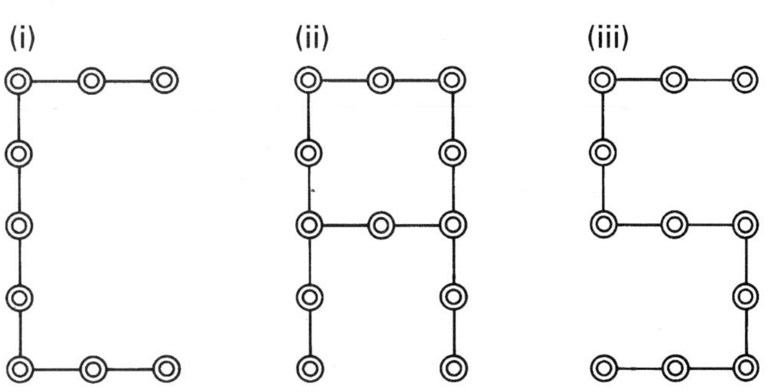

5. Four people of masses 60 kg, 65 kg, 62 kg and 75 kg sit on the four seats of the fairground ride shown below. The seats and the connecting arms are light. Find the radius of the circle described by the centre of mass when the ride rotates about O.

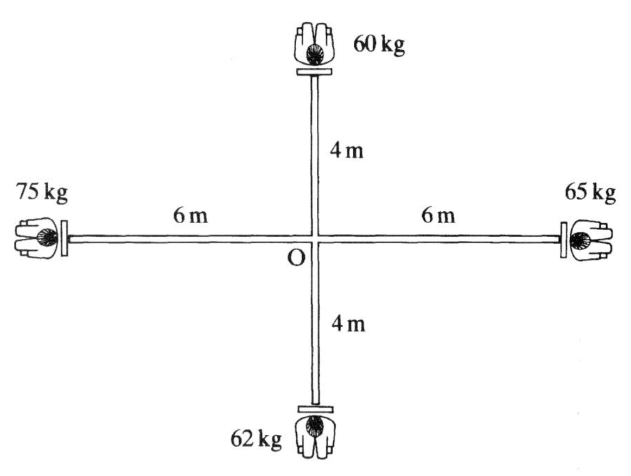

6. The following shapes are made out of uniform card. For each shape

(i) define an origin and x and y axes;

(ii) find the coordinates of the centre of mass.

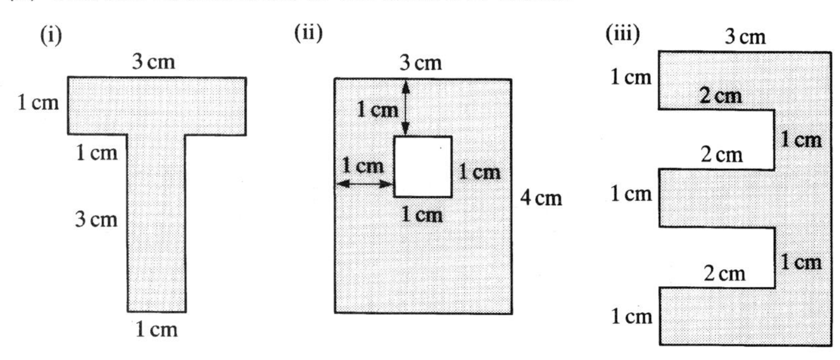

Exercise 3B continued

7. Uniform wooden bricks have length 20 cm and height 5 cm. They are glued together as shown in the diagram with each brick 5 cm to the right of the one below it. The origin is taken to be at O.

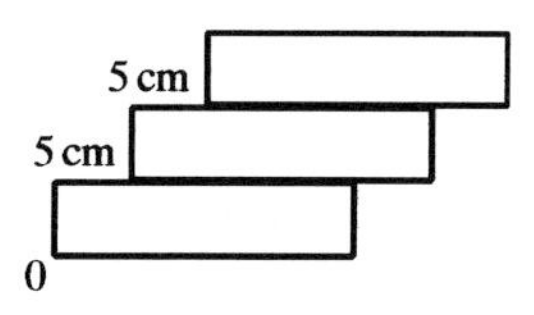

(i) Find the coordinates of the centre of mass for
(a) 1 (b) 2 (c) 3 (d) 4 (e) 5 bricks.

(ii) How many bricks is it possible to assemble in this way without their tipping over?

(iii) If the displacement were changed from 5 cm to 2 cm, how many bricks could then be assembled?

(iv) If the displacement is $\frac{1}{2}$ cm, what is the maximum height possible for the centre of mass of such an assembly of bricks without their tipping over?

8. A filing cabinet has the dimensions shown in the diagram. The body of the cabinet has mass 20 kg and its construction is such that its centre of mass is at a height of 60 cm, and is 25 cm from the back of the cabinet. The mass of a drawer and its contents may be taken to be 10 kg and its centre of mass to be 10 cm above its base and 30 cm from its front face.

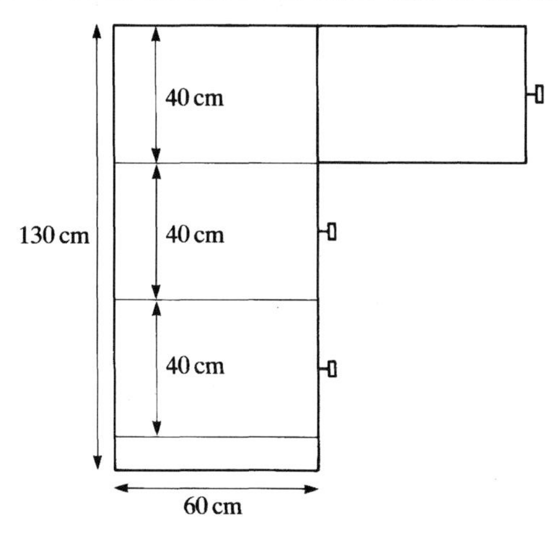

(i) Find the position of the centre of mass when all the drawers are closed.

(ii) Find the position of the centre of mass when the top two drawers are open.

(iii) Show that when all three drawers are fully opened the filing cabinet will tip over.

(iv) Two drawers are fully open. How far can the third one be opened without the cabinet tipping over?

9. A uniform rectangular lamina, ABCD, where AB is of length a and BC of length $2a$, has a mass $10m$. Further point masses m, $2m$, $3m$ and $4m$ are fixed to the points A, B, C and D, respectively.

Find the centre of mass of the system relative to x and y axes along AB and AD respectively.

If the lamina is suspended from the point A find the angle that the diagonal AC makes with the vertical.

To what must the mass at point D be altered if this diagonal is to hang vertically? [MEI]

10.

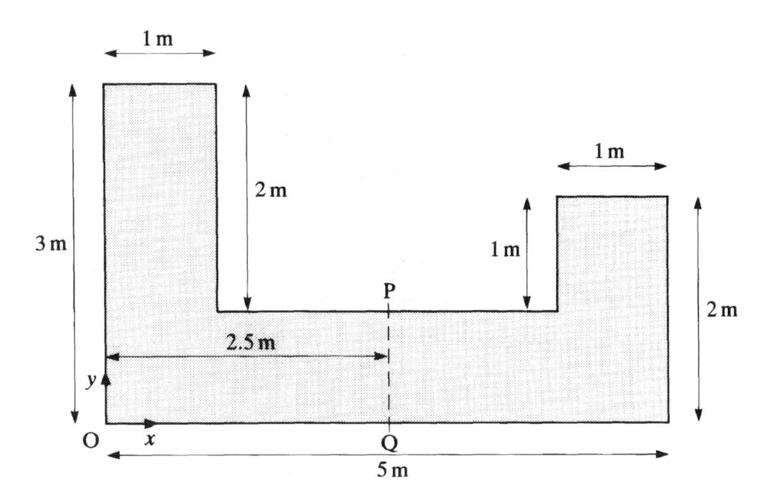

The diagram gives the dimensions of the design of a uniform metal plate. Using a co-ordinate system with O as origin, the x and y axes as shown and 1 metre as 1 unit,

(i) show that the centre of mass has y co-ordinate 1 and find its x co-ordinate.

The design requires the plate to have its centre of mass half way across (i.e. on the line PQ in the diagram), and in order to achieve this a circular hole centred on ($\frac{1}{2}$, $\frac{1}{2}$) is considered.

(ii) Find the appropriate radius for such a hole and explain why this idea is not feasible.

It is then decided to cut two circular holes each of radius r, both centred on the line $x = \frac{1}{2}$. The first hole is centered at ($\frac{1}{2}$, $\frac{1}{2}$) and the centre of mass of the plate is to be at P.

(iii) Find the value of r and the co-ordinates of the centre of the second hole.

[MEI]

Exercise 3B continued

11. The diagram shows how the human body can be modelled as a number of cylinders, and a sphere for the head. The table gives a typical mass and dimensions for each part.

	% mass	height/length	radius
head	6%	–	10 cm
body	52%	62 cm	15 cm
arms	6% each	70 cm	4 cm
legs	15% each	90 cm	7 cm

Find the centre of mass for the human in each of the positions shown below.

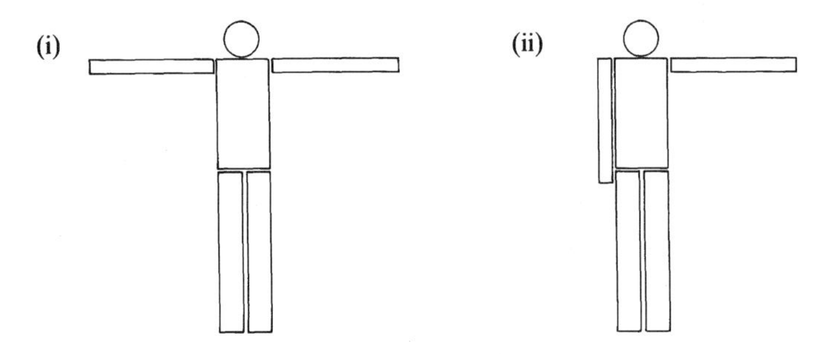

12. A drink can is cylindrical, with height h cm, and when empty its mass is m g. The drink that fills it has mass M g and can be taken to fill the can completely.

(i) Find the height of the centre of mass when the can is standing on a level table and

(a) is half full (b) a proportion, α, of the drink remains.

When the can is full the centre of mass is clearly half way up it, at height $\frac{1}{2}h$. The same is true when it is completely empty. In between these two extremes, the centre of mass is below the middle.

Show that when the centre of mass is at its minimum height

(i) $m\alpha^2 + 2\alpha m - m = 0$

(ii) the centre of mass lies on the surface of the drink.

Investigations

Mobile

Design and make a mobile whose pieces are circles of different radii.

Sign

Cut a piece of plywood into an irregular shape to make a novel pub sign. (All your angles should be 90°).

(i) Find the position of the centre of mass both by calculation and by experiment and compare your two answers.

(ii) The sign is hinged to the wall and supported by a stay wire running from a point on the wall above the hinge to a point on the sign. Find the tension in the wire both by calculation and by experiment and compare your values.

Bridge

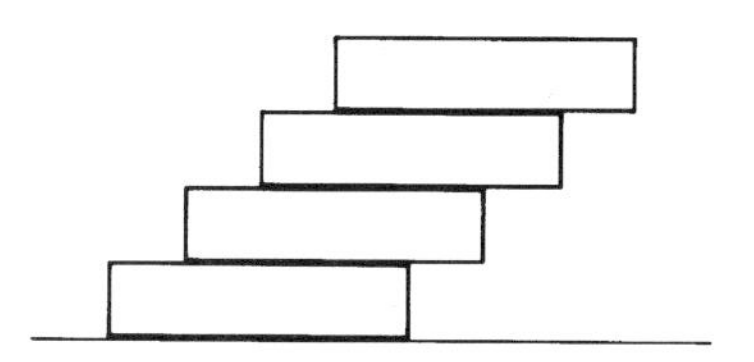

A bridge is made by placing identical bricks on top of each other as shown in the diagram. No glue or cement is used. How far can the bridge be extended without toppling over? You may use as many bricks as you like but only one is allowed at each level.

Finding the centre of mass

Collect a number of flat (but not necessarily uniform) objects, and investigate, for each of them, which is the most accurate method of determining its centre of mass.

(i) Calculation.

(ii) Balancing it on a pin.

(iii) Hanging it from 2 (or more) corners.

(iv) Balancing it on the edge of a table in a number of different orientations.

KEY POINTS

- The centre of mass of a body has the property that:
  the moment, about any point, of the whole mass of the body taken at the centre of mass is equal to the sum of the moments of the various particles comprising the body.

$$M\bar{r} = \sum_i m_i r_i \text{ where } M = \sum_i m_i$$

- In one dimension

$$M\bar{x} = \sum_i m_i$$

- In two dimensions

$$M\begin{pmatrix} \bar{x} \\ \bar{y} \end{pmatrix} = \sum_i m_i \begin{pmatrix} x_i \\ y_i \end{pmatrix}$$

- In three dimensions

$$M\begin{pmatrix} \bar{x} \\ \bar{y} \\ \bar{z} \end{pmatrix} = \sum_i m_i \begin{pmatrix} x_i \\ y_i \\ z_i \end{pmatrix}$$

4 Energy, work and power

I like work: it fascinates me. I can sit and look at it for hours.

Jerome K. Jerome.

This is a picture of a perpetual motion machine. What does this term mean and will this one work?

Energy

This chapter is about energy. In everyday life you encounter many forms of energy, such as heat, light, electricity and sound. You are familiar with the conversion of energy from one form to another: from chemical energy stored in coal to heat energy when you burn a coal fire; from electrical energy into the energy of a train's motion, and so on.

In any situation, energy is conserved. That means that the total energy, in all its forms, is constant. Energy can neither be created (as seems to be happening in the picture above) nor destroyed.

While the word 'energy' is often used loosely in ordinary speech, its technical use within the vocabulary of science must of course be precise. As a first step, the concept of energy, along with several other important ideas, is developed from Newton's Second Law in the following example.

For Discussion

A car of mass m kg is travelling at u ms^{-1} when the driver applies the accelerator for a period of t seconds. During this time the engine produces a constant driving force of F newtons. The ground is level and the road is straight and air resistance can be ignored.

The application of the driving force will have an effect on the speed of the car. So far we have used Newton's Second Law to model a situation like this, but now we need to look at it in a different way. What information is given by:

(i) the driving force $F \times$ the time t for which it is applied
(ii) the driving force $F \times$ the distance s over which it is applied?

The answers to these questions involve some very important concepts in mechanics: *impulse* and *work*, *momentum* and *energy*.

The first two relate to the forces acting on a body and the last two relate to the motion of the body at a particular instant.

- Impulse of force = force × time for which it acts
- Work done by force = force × distance moved in the direction of the force

} For a constant force

- Momentum of body = mass × velocity
- Kinetic energy of body = $\frac{1}{2}$ × mass × (speed)^2

Treating the car as a particle and applying Newton's Second Law:

$$F = ma$$

$$\Rightarrow a = \frac{F}{m}.$$

Since F is assumed constant, the acceleration is constant also, and the appropriate formulae may be used.

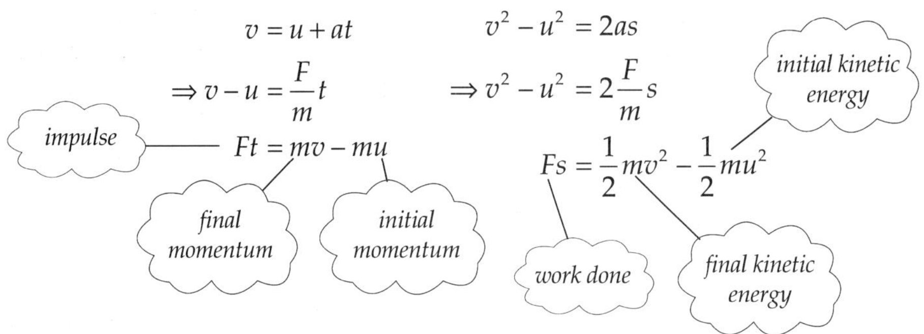

The two results can thus be written

- impulse = final momentum − initial momentum
- work done = final kinetic energy − initial kinetic energy

The first equation will be met again in Chapter 5. The quantities in the second equation, energy, work and also power, are the subject of this chapter.

Mechanical energy

In mechanics two forms of energy are particularly important.

Kinetic Energy

This is the energy which a body possesses because of its motion. The kinetic energy of a body of mass m moving with speed v is given by $\frac{1}{2}mv^2$.

Potential Energy

This is the energy which a body possesses because of its position. It may be thought of as stored energy which can be converted into kinetic or other forms of energy. Common forms of potential energy are *gravitational* potential energy (that possessed by a body in a high place) and *elastic* potential energy (that possessed by a stretched elastic string).

Both of these are forms of *mechanical energy*. In many, but by no means all, situations, mechanical energy is conserved. When it is not conserved, some of it is converted into other forms, like heat and sound.

Energy is a scalar quantity not a vector, having magnitude but no direction. The SI unit for energy is the joule (abbreviation J).

Work

When a body moves a distance s along the line of a force F, an amount of work given by Fs is done by the force, as in the following examples.

- A sledge, initially at rest, is pulled across smooth ice by a horizontal string in which there is a constant tension 40 N. When the sledge has travelled 18 m, the work done on it is

$$40 \times 18 = 720 \text{ joules.}$$

  Since the ice is smooth, this is all converted into kinetic energy. The sledge ends up with kinetic energy 720 joules. Thus if the mass of the sledge and its load is 40 kg, its final speed will be 6 ms^{-1} since $\frac{1}{2} \times 40 \times 6^2 = 720$.

- A train travelling on level ground is subject to a resisting force (from the brakes and air resistance) of 250 000 N for a distance of 5000 m. The forward force is thus − 250 000 N and the work done by it is

$$-250\,000 \times 5000 = -1\,250\,000\,000 \text{ joules.}$$

  This means that −1 250 000 000 joules of kinetic energy are gained by the train, in other words that 1 250 000 000 joules of kinetic energy are lost. These are converted to other forms of energy such as heat and perhaps a little sound.

- A brick, initially at rest is raised by a force averaging 40 N to a height 5 m above the ground where it is left stationary. The work done by the force on the brick is

$$40 \times 5 = 200 \text{ joules.}$$

It is important to realise that:

- work is done by a force
- work is only done when there is movement
- only movement that has a component along the line of the force contributes to the work done.

It is however quite common to speak about the work done by a person, say in pushing a lawn mower. It would be more precise to refer to the work done by the force produced by the person, but also rather long-winded.

Notice that if you stand holding a brick stationary above your head, painful though it may be, the force you are exerting on it is doing no work. Nor is this vertical force doing any work if you walk round the room keeping the brick at the same height. However once you start climbing the stairs, a component of the brick's movement is in the direction of the upward force that you are exerting on it, so the force is now doing some work.

The work-energy principal

The examples above illustrate the *work-energy principle* which states that:

The total work done by the forces acting on a body is equal to the increase in the kinetic energy of the body.

When applying the work-energy principle, you have to be careful to include *all* the forces acting on the body. In the example of a brick of weight 40 N being raised 5 metres vertically, starting and ending at rest, the change in kinetic energy is clearly

$$0 \text{ joules} - 0 \text{ joules} = 0 \text{ joules}$$

This seems paradoxical when it is clear that the force which raised the brick has done $40 \times 5 = 200$ joules of work. However the brick was subject to another force, namely its weight, which did $-40 \times 5 = -200$ joules of work on it, giving a total of $200 + (-200) = 0$ joules.

Conservation of mechanical energy

If a body is not subject to any forces it will continue to move in a straight line at constant speed, or will remain at rest. In such cases its kinetic energy clearly remains unaltered, and since it is subject to no forces, so does its potential energy. Mechanical energy is conserved.

When a body is subject to external forces, mechanical energy may or may not be conserved according to the nature of those forces. Forces under which mechanical energy is conserved are called conservative. Movement of a body against a conservative force gives it potential energy which is available to be converted at some subsequent time into kinetic energy. The commonest example of a conservative force is the force of gravity.

When a moving body is subject to a non-conservative or dissipative force like friction or air resistance, mechanical energy is not conserved. It is converted into other forms, like heat and sound.

EXAMPLE

The combined mass of a cyclist and her bicycle is 65 kg. She accelerated from rest to 8 ms^{-1} in 80 metres along a horizontal road.

(i) Calculate the work done by the net force in accelerating the cyclist and her bicycle.

(ii) Hence calculate the net forward force (assuming the force to be constant).

Solution

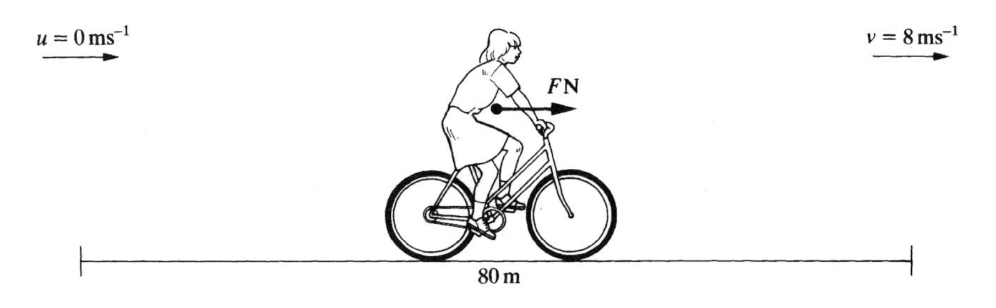

(i) The work done by the net force F is given by

$$\begin{aligned} \text{work} &= \text{change in kinetic energy} \\ &= \tfrac{1}{2}mv^2 - \tfrac{1}{2}mu^2 \\ &= \tfrac{1}{2} \times 65 \times 8^2 - 0 \\ &= 2080\text{ J} \end{aligned}$$

The work done is 2080 joules

(ii)

$$\begin{aligned} \text{Work done} &= Fs \\ &= F \times 80 \\ \text{So } 80F &= 2080 \\ F &= 26 \end{aligned}$$

The net forward force is 26 N.

EXAMPLE

A bullet of mass 25 g is fired at a wooden barrier 3 cm thick. When it hits the barrier it is travelling at 200 ms^{-1}. The barrier exerts a constant resistive force of 5000 N on the bullet.

(i) Does the bullet pass through the barrier and if so with what speed does it emerge?

(ii) Is energy conserved in this situation?

Solution

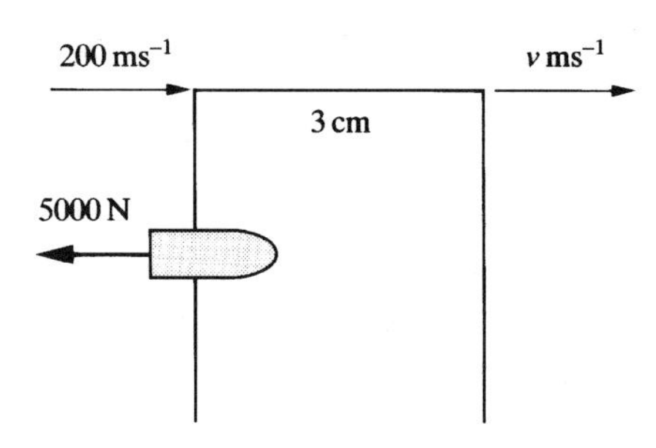

(i) The work done *by* the force is defined as the product of the force and the distance moved *in the direction of the force*. Since the bullet is moving in the direction opposite to the net resistive force, the work done by this force is negative.

$$\text{Work done} = -5000 \times 0.03\text{ J}$$
$$= -150\text{ J}$$

The initial kinetic energy of the bullet is

$$\text{Initial K.E.} = \tfrac{1}{2}mu^2$$
$$= \tfrac{1}{2} \times 0.025 \times 200^2$$
$$= 500\text{ J}$$

A loss in energy of 150 joules will not reduce kinetic energy to zero, so the bullet will still be moving on exit.
Since the work done is equal to the change in kinetic energy,

$$-150 = \tfrac{1}{2}mv^2 - 500$$

Solving for v

$$\tfrac{1}{2}mv^2 = 500 - 150$$
$$v^2 = \frac{2 \times (500 - 150)}{0.025}$$
$$v = 167$$

So the bullet emerges from the barrier with a speed of 167 ms^{-1}.

(ii) Total energy is conserved but there is a loss of mechanical energy of $\tfrac{1}{2}mu^2 - \tfrac{1}{2}mv^2 = 150$ joules. This energy is converted into non-mechanical forms such as heat and sound.

EXAMPLE

An aircraft of mass m kg is flying at a constant velocity $v\text{ ms}^{-1}$ horizontally. Its engines are providing a horizontal driving force F N.

(i) Draw a diagram showing the driving force, the lift force L N, the air resistance R N and the weight of the aircraft.

(ii) State which of these forces are equal in magnitude.

(iii) State which of the forces are doing no work.

(iv) In the case when $m = 100\,000$, $v = 270$ and $F = 350\,000$, find the work done in a 10 second period by those forces which are doing work, and show that the work-energy principle holds in this case.
At a later time the pilot increases the thrust of the aircraft's engines to 400 000 N. When the aircraft has travelled a distance of 30 km, its speed has increased to 300 ms^{-1}.

(v) Find the work done against air resistance during this period, and the average resistance force.

Solution

(i)

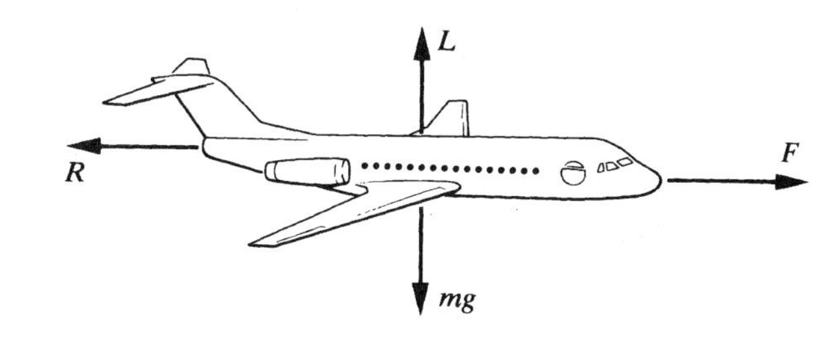

(ii) Since the aircraft is travelling at constant velocity it is in equilibrium.

Horizontal forces: $F = R$

Vertical forces: $L = mg$

(iii) Since the aircraft's velocity has no vertical component, the vertical forces, L and mg, are doing no work.

(iv) In 10 seconds at 270 ms^{-1} the aircraft travels 2700 m.

Work done by force $F = 350\,000 \times 2700 = 94\,500\,000$ J

Work done by force $R = 350\,000 \times -2700 = -94\,500\,000$ J

The work-energy principle states that in this situation

Work done by F + work done by R = Change in kinetic energy.

Now work done by F + work done by R $= (94\,500\,000 - 94\,500\,000) = 0$ J, and change in kinetic energy = 0 (since velocity is constant), so the work-energy principle does indeed hold in this case.

(v)

$$\begin{aligned}
\text{Total work done} &= \text{K.E. gained} \\
&= \tfrac{1}{2}mv^2 - \tfrac{1}{2}mu^2 \\
&= (\tfrac{1}{2} \times 100\,000 \times 300^2 - \tfrac{1}{2} \times 100\,000 \times 270^2) \\
&= 855 \times 10^6 \text{ J} \\
\text{Work done by driving force} &= 400\,000 \times 30\,000 \\
&= 12\,000 \times 10^6 \text{ J} \\
\text{Work done by resistance force} &= (855 \times 10^6 - 12\,000 \times 10^6) \\
&= -11\,145 \times 10^6 \text{ J} \\
\text{Average force} \times \text{distance} &= \text{work done by force} \\
\text{average force} \times 30\,000 &= -11\,145 \times 10^6
\end{aligned}$$

$\Rightarrow$ The average resistance force is 371 500 N (in the negative direction.)

NOTE

When an aircraft is in flight, most of the work done by the resistance force results in air currents and the generation of heat. A typical large jet cruising at 35 000 feet has a body temperature about 30° C above the surrounding air temperature. For supersonic flight the temperature difference is much greater. Concorde flies with a skin temperature more than 200° C above the surrounding air.

Exercise 4A

1. Find the kinetic energy of the following objects.

(i) An ice skater of mass 50 kg travelling with speed $10\ \mathrm{ms}^{-1}$.

(ii) An elephant of mass 5 tonnes moving at $4\ \mathrm{ms}^{-1}$.

(iii) A train of mass 7000 tonnes travelling at $40\ \mathrm{ms}^{-1}$.

(iv) The Moon, mass 7.4×10^{22} kg, travelling at $1000\ \mathrm{ms}^{-1}$ in its orbit round the earth.

(v) A bacterium of mass 2×10^{-16} g which has speed $1\ \mathrm{mms}^{-1}$.

2. Find the work done by a man in the following situations.

(i) He pushes a packing case of mass 35 kg a distance of 5 m across a rough floor against a resistance force of 200 N. The case starts and finishes at rest.

(ii) He pushes a packing case of mass 35 kg a distance of 5 m across a rough floor against a resistance force of 200 N. The case starts at rest and finishes with a speed of $2\ \mathrm{ms}^{-1}$.

(iii) He pushes a packing case of mass 35 kg a distance of 5 m across a rough floor against a resistance force of 200 N. Initially the case has speed $2\ \mathrm{ms}^{-1}$ but it ends at rest.

(iv) He is handed a packing case of mass 35 kg. He holds it stationary, at the same height, for 20 s and then someone else takes it from him.

3. A sprinter of mass 60 kg is at rest at the beginning of a race and accelerates to $12\ \mathrm{ms}^{-1}$ in a distance of 30 m. Assume air resistance to be negligible.

(i) Calculate the kinetic energy of the sprinter at the end of the 30 m.

(ii) Write down the work done by the sprinter over this distance.

(iii) Calculate the forward force exerted by the sprinter, assuming it to be constant, using work = force $\times$ distance.

(iv) Using force = mass $\times$ acceleration and the constant acceleration formulae, show that this force is consistent with the sprinter having speed $12\ \mathrm{ms}^{-1}$ after 30 m.

4. A sports car of mass 1.2 tonnes accelerates from rest to $30\ \mathrm{ms}^{-1}$ in a distance of 150 m. Assume air resistance to be negligible.

(i) Calculate the work done in accelerating the car. Does your answer depend on an assumption that the driving force is constant?

(ii) If the driving force is in fact constant, what is its magnitude?

5. A car of mass 1600 kg is travelling at speed $25\ \mathrm{ms}^{-1}$ when the brakes are applied so that it stops after moving a further 75 m.

(i) Find the work done by the brakes.

(ii) Find the retarding force from the brakes, assuming that it is constant and that other resistive forces may be neglected.

6. A bullet of mass 20 g, found at the scene of a police investigation, had penetrated 16 cm into a wooden post. The speed for that type of bullet is known to be 80 ms^{-1}.

(i) Find the kinetic energy of the bullet before it entered the post.

(ii) What happened to this energy when the bullet entered the wooden post?

(iii) Write down the work done in stopping the bullet.

(iv) Calculate the resistive force on the bullet, assuming it to be constant.

Another bullet of the same mass and shape had clearly been fired from a different and unknown type of gun. This bullet had penetrated 20 cm into the post.

(v) Estimate the speed of this bullet before it hit the post.

7. The Highway Code gives the braking distance for a car travelling at 22 ms^{-1} (50 mph) to be 38 m (125 ft). A car of mass 1300 kg is brought to rest in just this distance. It may be assumed that the only resistance forces come from the car's brakes.

(i) Find the work done by the brakes.

(ii) Find the average force exerted by the brakes.

(iii) What happened to the kinetic energy of the car?

(iv) What happens when you drive a car with the handbrake on?

8. A car of mass 1200 kg experiences a constant resistance force of 600 N. The driving force from the engine depends upon the gear, as shown in the table.

Gear	1	2	3	4
Force (N)	2800	2100	1400	1000

The car is driven for 20 m in first gear, 40 m in second, 80 m in third and 100 m in fourth. How fast is the car travelling at the end?

9. In this question take g to be 10 ms^{-2}. A chest of mass 60 kg is resting on a rough horizontal floor. The coefficient of friction between the floor and the chest is 0.4. A woman pushes the chest in such a way that its speed-time graph is as shown below.

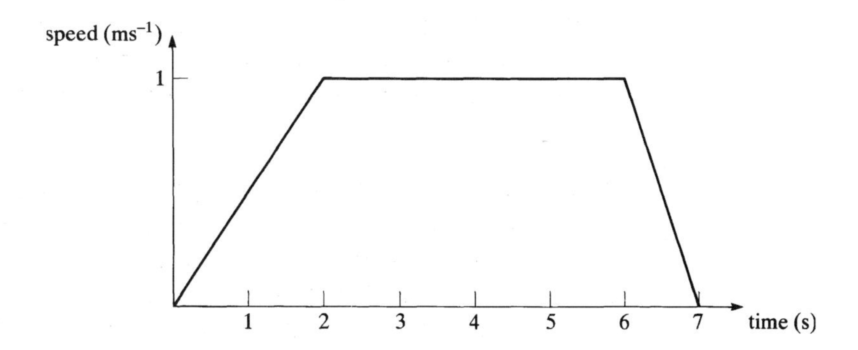

Exercise 4A continued

(i) Find the force of frictional resistance acting on the chest when it moves.

(ii) Use the speed-time graph to find the total distance travelled by the chest.

(iii) Find the total work done by the woman.

(iv) Find the acceleration of the chest in the first 2 seconds of its motion and hence the force exerted by the woman during this time, and the work done.

(v) In the same way find the work done by the woman during the time intervals 2 to 6 seconds, and 6 to 7 seconds.

(vi) Show that your answers to parts (iv) and (v) are consistent with your answer to part (iii).

10. A ball of mass 0.25 kg was dropped from a height of 2 m onto a concrete floor. Assume that the only force acting on the ball was that from gravity, and take g to be 10 ms^{-2}.

(i) Calculate the speed of the ball just before it bounced.

The ball rebounded to a maximum height of 1.8 m.

(ii) Find the work done by the force of gravity in slowing the ball down.

(iii) Calculate the speed of the ball as it left the floor.

(iv) Calculate the loss of kinetic energy during the bounce. What happened to this energy?

Gravitational potential energy

As you have seen, kinetic energy (K.E.) is the energy that an object has because of its motion. Potential energy (P.E.) is the energy an object has because of its position. The units of potential energy are the same as those of kinetic energy or any other form of energy, namely joules.

One form of potential energy is *gravitational potential energy*. The gravitational potential energy of the object in figure 4.1 of mass m kg at height h metres above a fixed reference level, O, is mgh joules. If it falls to the reference level, gravity does mgh joules of work and the body loses mgh joules of potential energy.

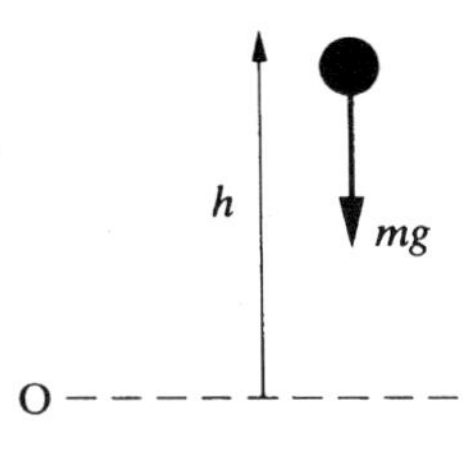

Figure 4.1

The work done by gravity is often thought of as a loss of gravitational potential energy.

If a mass m kg is *raised* through a distance h metres, the gravitational potential energy *increases* by mgh joules. If a mass m kg is *lowered* through a distance h metres the gravitational potential energy *decreases* by mgh joules.

EXAMPLE

Calculate the gravitational potential energy, relative to the ground, of a ball of mass 0.15 kg at a height of 2 m above the ground.

Solution

Mass $m = 0.15$, height $h = 2$.

Gravitational potential energy $= mgh$
$= 0.15 \times 9.8 \times 2$
$= 2.94$ joules.

NOTE

If the ball falls:

loss in P.E. = work done by gravity
= gain in K.E.

There is no change in the total energy (P.E. + K.E.) of the ball.

Using conservation of mechanical energy

When gravity is the only force which does work on a body, mechanical energy is conserved. When this is the case, many problems are easily solved using energy.

EXAMPLE

A skier slides down a smooth ski slope which is at an angle of 30° to the horizontal. Find the speed of the skier when he reaches the bottom of the slope.

At the foot of the slope the ground becomes horizontal and is made rough in order to help him to stop. The coefficient of friction between his skis and the ground is $\frac{1}{4}$.

(i) Find how far the skier travels before coming to rest.
(ii) In what way is your model unrealistic?

Solution

The skier is modelled as a particle.

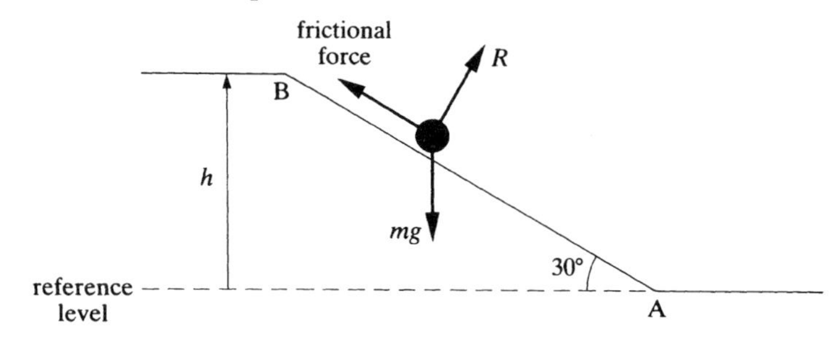

(i) Since in this case the slope is smooth, the frictional force is zero. The skier is subject to two external forces, his weight mg and the normal reaction from the slope.
The normal reaction between the skier and the slope does no work because the skier does not move in the direction of this force. The only force which does work is gravity, so mechanical energy is conserved.

$$\begin{aligned} \text{Total mechanical energy at B} &= mgh + \tfrac{1}{2}mu^2 \\ &= (m \times 9.8 \times 400 \sin 30° + 0) \\ &= 1960\,m\,\text{J} \end{aligned}$$

$$\text{Total mechanical energy at A} = (0 + \tfrac{1}{2}mv^2)\,\text{J}$$

Since mechanical energy is conserved,

$$\begin{aligned} \tfrac{1}{2}mv^2 &= 1960\,m \\ v^2 &= 3920 \\ v &= 62.6 \end{aligned}$$

The skier's speed at the bottom of the slope is 62.6 ms^{-1}.

Notice that the mass of the skier cancels out. Using this model, all skiers should arrive at the bottom of the slope with the same speed.)

For the horizontal part there is some friction. Suppose that the skier travels a further distance s before stopping.

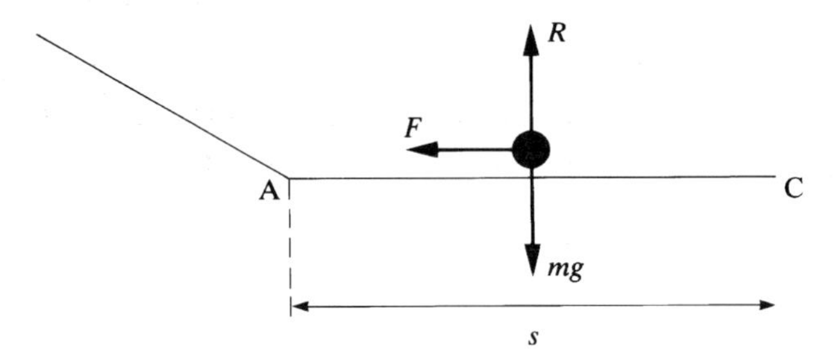

Coulomb's law of friction gives $\quad F = \mu R = \tfrac{1}{4}R.$

Since there is no vertical acceleration we can also say

$$R = mg$$

So $\quad F = \tfrac{1}{4}mg.$

Work done by the friction force $F \times (-s) = \tfrac{1}{4}mgs.$

negative because the motion is in the opposite direction to the force

The increases in kinetic energy between A and C = $(0 - \tfrac{1}{2}mv^2)$ joules.
Using the work-energy principle

$$-\tfrac{1}{4}mgs = -\tfrac{1}{2}mv^2 = -1960\,m.$$

Solving for s gives $s = 800$.

So the distance the skier travels before stopping is 800 metres.

(ii) The assumptions made in solving this problem are that friction on the slope and air resistance are negligible, and that the slope ends in a smooth curve at A. Clearly the speed of 62.6 ms^{-1} is very high, so our assumption that friction and air resistance are negligible must be suspect.

EXAMPLE

Ama, whose mass is 40 kg, is taking part in an assault course. The obstacle shown in the picture is a river at the bottom of a ravine 8 m wide which she has to cross by swinging on a rope 5 m long secured to a point on the branch of a tree, immediately above the centre of the ravine.

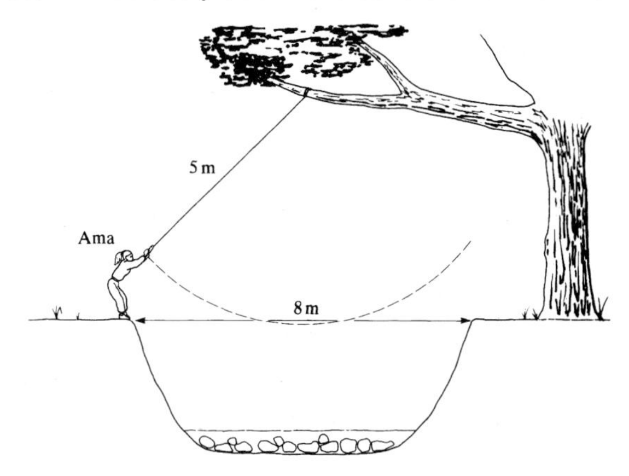

(i) Find how fast Ama is travelling at the lowest point of her crossing.
(ii) If Ama launches herself off at a speed of 1 ms^{-1} (in the right direction) will her speed be 1 ms^{-1} faster throughout her crossing?

Solution

(i)

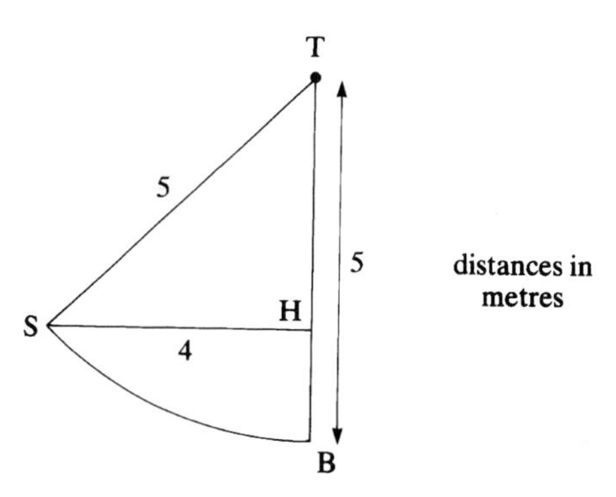

The vertical height Ama loses is HB in the diagram

Using Pythagoras

$$TH = \sqrt{5^2 - 4^2} = 3$$

$$HB = 5 - 3 = 2$$

$$\text{P.E. lost} = mgh$$

$$\text{K.E. gained} = \tfrac{1}{2}mv^2 - 0$$

$$= \tfrac{1}{2} \times 40 \times v^2$$

By conservation of energy, K.E. gained $= $ P.E. lost

$$\tfrac{1}{2}\times 40\times v^2 = 40\times 9.8\times 2$$

$$v = 6.26$$

Ama is travelling at $6.26\,\text{ms}^{-1}$.

(ii) If she has initial speed $1\,\text{ms}^{-1}$ at S and speed $v\,\text{ms}^{-1}$ at B, her initial K.E. is $\frac{1}{2}\times 40\times 1^2$ joules and her K.E. at B is $\frac{1}{2}\times 40\times v^2$.

Using conservation of energy,

$$\tfrac{1}{2}\times 40\times v^2 - \tfrac{1}{2}\times 40\times 1^2 = 40\times 9.8\times 2$$

This gives $v = 6.34$, so Ama's speed at the lowest point is now $6.34\,\text{ms}^{-1}$, only $0.08\,\text{ms}^{-1}$ faster than in part (i), so she clearly will not travel $1\,\text{ms}^{-1}$ faster throughout.

HISTORICAL NOTE

James Joule was born in Salford in Lancashire on Christmas Eve 1818. He studied at Manchester University at the same time as the famous chemist, Dalton.

Joule spent much of his life conducting experiments to measure the equivalence of heat and mechanical forms of energy to ever increasing degrees of accuracy. Working with Thompson, he also made the discovery that when a gas is allowed to expand without doing work against external forces it cools. It was this discovery that paved the way for the development of refrigerators.

Joule died in 1889 but his contribution to science is remembered with the SI Unit for energy named after him.

Work and kinetic energy for two dimensional motion

For Discussion

Imagine that you are cycling along a level winding road in a strong wind. Suppose that the strength and direction of the wind are constant, but because the road is winding sometimes the wind is directly against you but at other times it is from your side.

Discuss how the work you do in travelling a certain distance – say 1 m – changes with your direction.

Work done by a force at an angle to the direction of motion

You have probably deduced that as a cyclist you would do work against the component of the wind force that is directly against you. The sideways component does not resist your forward progress.

Suppose that you are sailing and the angle between the direction in which the wind is blowing and the direction of your motion is θ. The force of the wind on your boat is F and in a certain time you travel a distance d in the direction of F, see figure 4.2. During that time you actually travel a distance s, along the line OP.

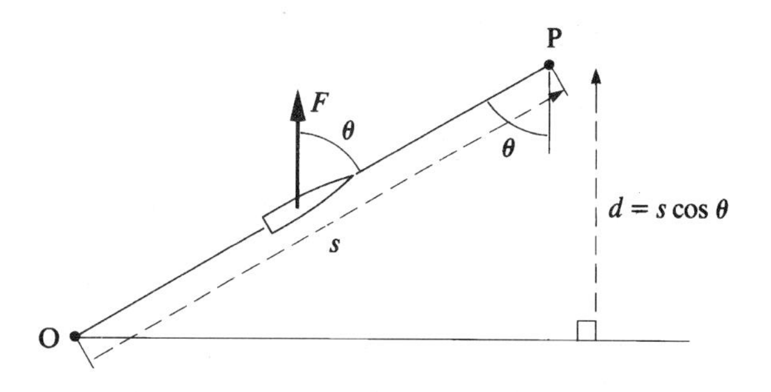

Figure 4.2

Work done by $F = Fd$

Since $d = s\cos\theta$, the work done by the force F is $Fs\cos\theta$. This can also be written as the product of the component of F along OP, $F\cos\theta$, and the distance moved along OP, s.

$$F \times s\cos\theta = F\cos\theta \times s$$

[Notice that the direction of F is that of the force of the wind on your boat. This is not necessarily the same as the direction of the wind, it depends on how you have set your sails].

EXAMPLE

As a car of mass m drives up a slope at an angle α to the horizontal it experiences a constant resistive force F and a driving force T. What can be deduced about the work done by T as the car moves a distance d uphill if:

(i) the car moves at constant speed?

(ii) the car slows down?

(iii) the car gains speed?

Solution

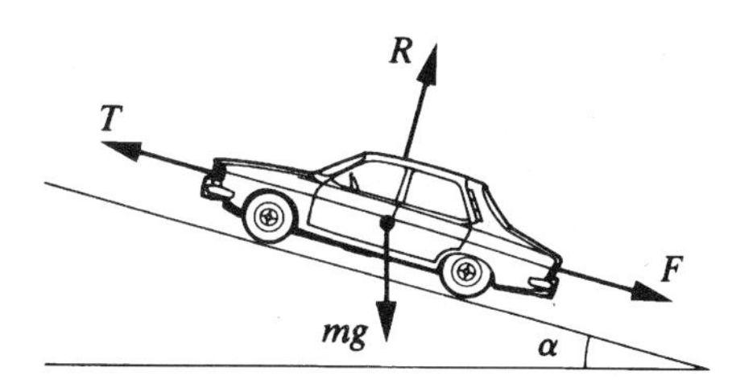

The diagram shows the forces acting on the car. The table shows the work done by each force. The normal reaction, R, does no work as the car moves no distance in the direction of R.

Force	Work done
Resistance F	$-Fd$
Normal reaction R	0
Force of gravity mg	$-mgd\cos(90-\alpha) = -mgd\sin\alpha$
Driving force T	Td
Total work done	$Td - Fd - mgd\sin\alpha$

(i) If the car moves at a constant speed there is no change in kinetic energy so the total work done is zero, giving

Work done by T is

$$Td = Fd + mgd\sin\alpha.$$

(ii) If the car slows down the total work done by the forces is negative, hence

Work done by T is

$$Td < Fd + mgd\sin\alpha.$$

(iii) If the car gains speed the total work done by the forces is positive

Work done by T is

$$Td > Fd + mgd\sin\alpha.$$

Exercise 4B

1. Calculate the gravitational potential energy, relative to the reference level OA, for each of the objects shown.

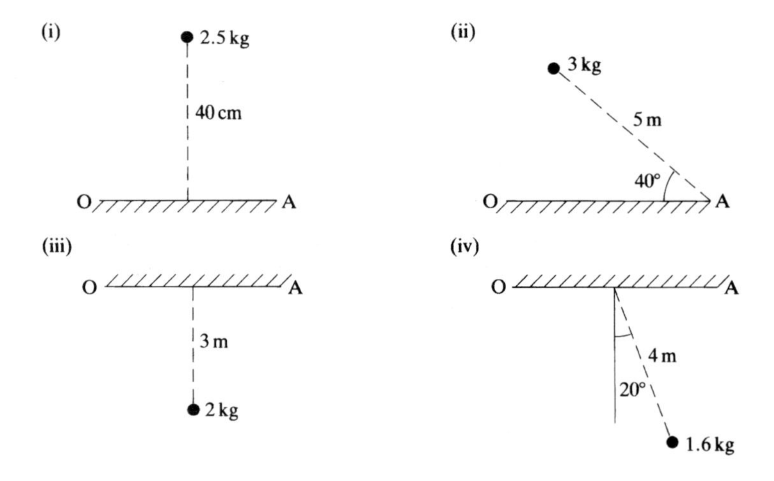

2. Calculate the change in gravitational potential energy when each object moves from A to B in the situations shown opposite. State whether the change is an increase or a decrease.

Exercise 4B continued

(i)

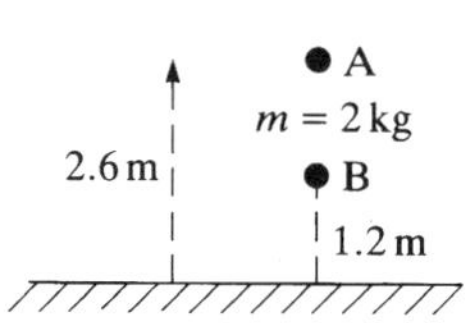

(ii)

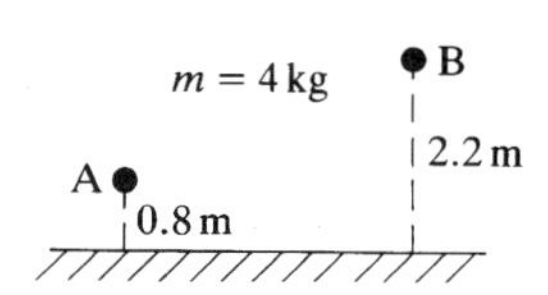

(iii)

1 m
A
3 m
m= 0.6 kg
B

3. A vase of mass 1.2 kg is lifted from ground level and placed on a shelf at a height of 1.5 m. Find the work done against the force of gravity.

4. Find the increase in gravitational potential energy of a woman of mass 60 kg who climbs to the 12th floor of a block of flats. The distance between floors is 3.3 m.

5. A car of mass 0.9 tonnes is driven 200 m up a slope inclined at $5°$ to the horizontal. There is a resistance force of 100 N.

(i) Find the work done by the car against gravity.

(ii) Find the work done against the resistance force.

(iii) When asked to work out the total work done by the car, a student replied '$(900g + 100) \times 200$ N'. Explain the error in this answer.

6. A sledge of mass 10 kg is being pulled across level ground by a rope which makes an angle of $20°$ with the horizontal. The tension in the rope is 80 N and there is a resistance force of 14 N. Initially the sledge is at rest.

(i) Find the work done by (a) the tension in the rope, (b) the resistance force while the sledge moves a distance of 20 m.

(ii) Find the speed of the sledge after it has moved 20 m.

7. A bricklayer carries a hod of bricks of mass 25 kg up a ladder of length 10 m inclined at an angle of $60°$ to the horizontal.

(i) Calculate the increase in the gravitational potential energy of the bricks.

(ii) If instead he had raised the bricks vertically to the same height, using a rope and pulleys, would the increase in potential energy be (a) less, (b) the same, or (c) more than in part (i)?

8. A girl of mass 45 kg slides down a smooth water chute of length 6 m inclined at an angle of $40°$ to the horizontal.

(i) Find (a) the decrease in her potential energy, (b) her speed at the bottom.

(ii) How are answers to part (i) affected if the slide is not smooth?

Exercise 4B continued

9. A gymnast of mass 50 kg swings on a rope of length 10 m. Initially the rope makes an angle of $60°$ with the horizontal.

(i) Find the decrease in her potential energy when the rope has reached the vertical.

(ii) Find her kinetic energy and hence her speed when the rope is vertical, assuming that air resistance may be neglected.

(iii) The gymnast continues to swing. What angle will the rope make with the vertical when she is next temporarily at rest?

(iv) Explain why the tension in the rope does no work.

10. A stone of mass 0.2 kg is dropped from the top of a building 78.4 m high. After t seconds it has fallen a distance x m and has speed v ms^{-1}.

(i) What is the gravitational potential energy of the stone relative to ground level when it is at the top of the building?

(ii) What is the potential energy of the stone t seconds later?

(iii) Show that, for certain values of t, $v^2 = 19.6\,x$ and state the range of values of t for which it is true.

(iv) Find the speed of the stone when it is half way to the ground.

(v) At what height will the stone have half its final speed?

11. Wesley, whose mass is 70 kg, inadvertently steps off a bridge 50 m above water. When he hits the water, Wesley is travelling at 25 ms^{-1}.

(i) Calculate the potential energy Wesley has lost and the kinetic energy he has gained.

(ii) Find the size of the resistance force acting on Wesley while he is in the air, assuming it to be constant.

Wesley descends to a depth of 5 m below the water surface, then returns to the surface.

(iii) Find the total upthrust (assumed constant) acting on him while he is moving downwards in the water.

12. A hockey ball of mass 0.15 kg is hit from the centre of a pitch. Its position vector, in metres, t seconds later is modelled by

$$\mathbf{r} = 10\,t\,\mathbf{i} + (\,10\,t - 4.9\,t^2\,)\,\mathbf{j}$$

where the unit vectors $\mathbf{i}$ and $\mathbf{j}$ are in directions along the line of the pitch and vertically upwards.

(i) What value of g is used in this model?

(ii) Find an expression for the gravitational potential energy of the ball at time t. For what values of t is your answer valid?

(iii) What is the maximum height of the ball? What is its velocity at that instant?

(iv) Find the initial velocity, speed and kinetic energy of the ball.

(v) Show that according to this model mechanical energy is conserved and state what modelling assumption is implied by this. Is it reasonable in this context?

13. A ski-run starts at altitude 2471 m and ends at 1863 m.

(i) If all resistance forces could be ignored, what would the speed of the skier be at the end of the run?

A particular skier of mass 70 kg actually attains a speed of $42\ \text{ms}^{-1}$. The length of the run is 3.1 km.

(ii) Find the average force of resistance acting on a skier.

Two skiers are equally skilful.

(iii) Which would you expect to be travelling faster by the end of the run, the heavier or the lighter?

14. A tennis ball of mass 0.06 kg is hit vertically upwards with speed $20\ \text{ms}^{-1}$ from a point 1.1 m above the ground. It reaches a height of 16 m.

(i) Find the initial kinetic energy of the ball, and its potential energy when it is at its highest point.

(ii) Calculate the loss of mechanical energy due to air resistance.

(iii) Find the magnitude of the air resistance force on the ball, assuming it to be constant while the ball is moving.

(iv) With what speed does the ball land?

15.

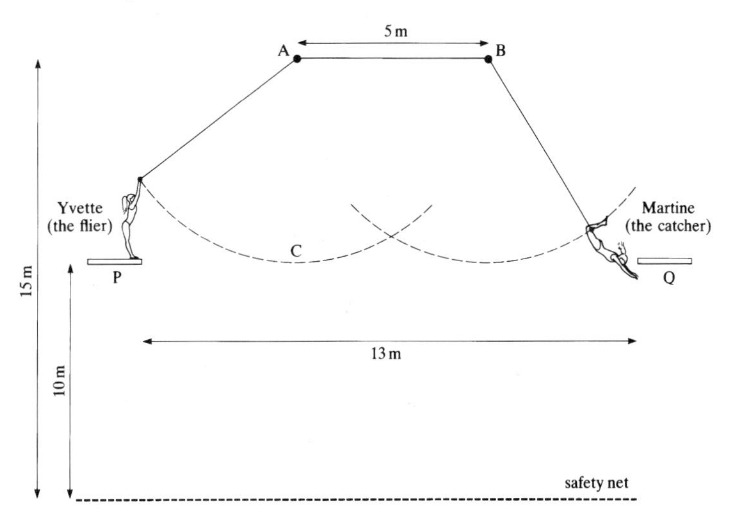

Yvette and Martine are trapeze artists in a circus. Their equipment is shown in the diagram. The trapezes are suspended from points A and B, 5 m apart but at the same height, 15 m above the safety net. The two platforms P and Q are at the same height, 10 m above the safety net, 13 m apart and placed symmetrically with respect to points A and B.

Yvette, who is the 'flier' of the team, holds her trapeze while standing on the edge of platform P, with her arms straight above her. In this position her hands are 2 m above the platform.

Exercise 4B continued

(i) How long is the rope of Yvette's trapeze?

(ii) How fast is Yvette travelling when she passes the lowest point (C) of her trapeze's arc? (Assume she does not push herself off).

As part of their act they frighten the audience. Yvette lets go of her trapeze at point C and Martine pretends to forget to catch her. Yvette falls into the safety net.

(iii) How long should the safety net be?

(iv) Describe, and comment on, the assumptions you have made in modelling this situation.

16. Akosua draws water from a well 12 m below the ground. Her bucket holds 5 kg of water and by the time she has pulled it to the top of the well it is travelling at $1.2\ \text{ms}^{-1}$.

(i) How much work does Akosua do in drawing the bucket of water?

On an average day 150 people in the village each draw 6 such buckets of water. One day a new electric pump is installed that takes water from the well and fills an overhead tank 5 m above ground level every morning. The flow rate through the pump is such that the water has speed $2\ \text{ms}^{-1}$ on arriving in the tank.

(ii) Assuming that the villagers' demand for water remains unaltered, how much work does the pump do in one day?

It takes the pump 1 hour to fill the tank each morning.

(iii) At what rate does the pump do work, in joules per second (Watts)?

Power

It is claimed that a motorcycle engine can develop a maximum *power* of 26.5 kW at a top *speed* of 103 mph. This suggests that power is related to speed and this is indeed the case.

Power is the rate at which work is done. A powerful car does work at a greater rate than a less powerful one. For a constant force, *F*, work is defined as *Fs* and we can differentiate this to find an expression for power.

$$\begin{aligned} \text{Power} &= \text{rate of doing work} \\ &= \frac{\mathrm{d}}{\mathrm{d}t}(Fs) \\ &= F\frac{\mathrm{d}s}{\mathrm{d}t} \,(F \text{ is constant}) \\ &= Fv. \end{aligned}$$

The power of a vehicle moving at speed v under a driving force F is given by Fv.

NOTE

This result is true whether F is constant or not. The use of constant F in the derivation above was for simplicity.

For a motor vehicle the power is produced by the engine, whereas for a bicycle it is produced by the cyclist. They both make the wheels turn, and the friction between the rotating wheels and the ground produces a forward force on the machine.

The unit of power is the watt (W), named after the physicist James Watt. The power produced by a force of 1 newton acting on an object that is moving at $1\,\mathrm{ms}^{-1}$ is 1 watt. Because the watt is such a small unit you will probably use kilowatts more often (1kW = 1000W).

EXAMPLE

A Kawasaki GPz 305 motorcycle has a maximum power output of 26.5 kW and a top speed of 103 mph ($46\,\mathrm{ms}^{-1}$). Find the force exerted by the motorcycle engine when the motorcycle is travelling at top speed.

Solution

If the force exerted by the motorcycle engine is F, then the expression for power, $P = Fv$, gives

$$\begin{aligned} F &= \frac{P}{v} \\ &= \frac{26500}{46} \\ &= 576 \end{aligned}$$

The force exerted by the engine at top speed is 576 N.

EXAMPLE

The power that a car is producing when moving at constant speed is 45 kW. If the car experiences a resistance of 1700 N, what is its speed?

Solution

As the car is travelling at a constant speed, there is no resultant force on the car. In this case the forward force of the engine must have the same magnitude as the resistance forces, i.e. 1700 N.

If the speed of the car is v ms^{-1} the expression for the power, $P = F v$, gives

$$v = \frac{P}{F}$$
$$= \frac{45000}{1700}$$
$$= 26.5$$

The speed of the car is 26.5 ms^{-1} (which is approximately 60 mph).

HISTORICAL NOTE

James Watt was born in 1736 in Greenock in Scotland, the son of a house- and ship-builder. As a boy James was frail and consequently for a long time was taught by his mother rather than going to school. This allowed him to spend time in his father's workshop where he developed practical and inventive skills.

As a young man he set himself up as a manufacturer of mathematical instruments: quadrants, scales, compasses and so on. One day he was repairing a model steam engine for a friend and noticed that its design was very wasteful of steam. He proposed an alternative arrangement, with a separate condenser, which was to become standard on later steam engines. This was the first of many engineering inventions which made possible the subsequent industrial revolution.

James Watt died in 1819, a well known and highly respected man. His name lives on as the SI unit for power.

Exercise 4C

1. A builder hoists bricks up to the top of the house he is building. Each brick weighs 3.5 kg and the house is 9 m high. In the course of one hour the builder raises 120 bricks from ground level to the top of the house, where they are unloaded by his mate.

(i) Find the increase in gravitational potential energy of one brick when it is raised in this way.

(ii) Find the total work done by the builder in one hour of raising bricks.

(iii) Find the average power with which he is working.

2. A weightlifter takes 2 seconds to lift 120 kg from the floor to a position 2 m above it where the weight has to be held stationary.

(i) Calculate the work done by the weightlifter.

(ii) Calculate the average power developed by the weightlifter.

The weightlifter is using the 'clean and jerk' technique. This means that in the first stage of the lift he raises the weight 0.8 m from the floor in 0.5 s. He then holds it stationary for 1 s before lifting it up to the final position in another 0.5 s.

(iii) Find the average power developed by the weightlifter during each of the stages of the lift.

3. A winch is used to pull a crate of mass 180 kg up a rough slope of angle 30° against a frictional force of 450 N. The crate moves at a steady speed, v, of 1.2 ms^{-1}.

(i) Calculate the gravitational potential energy given to the crate during 30 seconds.

(ii) Calculate the work done against friction during this time.

(iii) Calculate the total work done per second by the winch.

The cable from the winch to the packing case runs parallel to the slope.

(iv) Calculate the tension, T, in the cable.

(v) What information is given by $T \times v$?

4. The power output from the engine of a car which is travelling along level ground at a constant speed of 33 ms^{-1} is 23 200 watts. Find the total resistance on the car under these conditions.

5. A motorcyclist, of total mass (rider and cycle) 300 kg, is riding on level ground at a constant speed of 27 ms^{-1}. The power ouput of the engine is 16 500 W.

(i) Find the total resistance on the motorcyclist.

(ii) You were given one piece of unnecessary information. Which is it?

6. A crane is raising a load of 500 tonnes at a steady rate of 5 cms^{-1}. What power is the engine of the crane producing? (Assume that there are no forces from friction or air resistance).

7. A cyclist, travelling at a constant speed of 8 ms^{-1} along a level road, experiences a total resistance of 70 N.

(i) Find the power which the cyclist is producing.

(ii) Find the work done by the cyclist in 5 minutes under these conditions.

8. A conveyor belt picks up stationary sacks of grain and delivers them to a place 5 m higher with speed 1.5 ms^{-1}. The mass of one sack is 25 kg and they are delivered at the rate of one sack every 6 seconds.

(i) Calculate the total mechanical energy given to one sack by the conveyor belt.

(ii) Calculate the average power with which the conveyor's belt is working, assuming that frictional forces may be ignored.

9. A train consists of a diesel shunter of mass 100 tonnes pulling a truck of mass 25 tonnes along a level track. The engine is working at a rate of 125 kW. The resistance to motion of the truck and shunter is 50 N per tonne.

Exercise 4C continued

(i) Calculate the constant speed of the train.

While travelling at this constant speed, the truck becomes uncoupled. The shunter engine continues to produce the same power.

(ii) Find the acceleration of the shunter immediately after this happens.

(iii) Find the greatest speed the shunter can now reach.

10. A supertanker of mass 4×10^8 kg is steaming at a constant speed of $8\ \text{ms}^{-1}$. The resistance force is 2×10^6 N.

(i) What power are the ship's engines producing?

One of the ship's two engines suddenly fails but the other continues to work at the same rate.

(ii) Find the deceleration of the ship immediately after the failure.

The resistance force is directly proportional to the speed of the ship.

(iii) Find the eventual steady speed of the ship under one engine only, assuming that the single engine maintains constant power output.

11. A car has a maximum speed of $50\ \text{ms}^{-1}$ and a maximum power output of 40 kW. The resistance force, R N at speed $v\ \text{ms}^{-1}$ is modelled by

$$R = kv$$

(i) Find the value of k.

(ii) Find the resistance force when the car's speed is $20\ \text{ms}^{-1}$.

(iii) Find the power needed to travel at a constant speed of $20\ \text{ms}^{-1}$ along a level road.

Investigations

Crawler Lanes

Sometimes on single carriageway roads or even some motorways, crawler lanes are introduced for slow-moving, heavily laden lorries. Investigate how steep a slope can be before a crawler lane is needed.

Data: Typical power output for a large lorry : 45 kW

Typical mass of a large laden lorry : 32 tonnes.

Stimpmeter

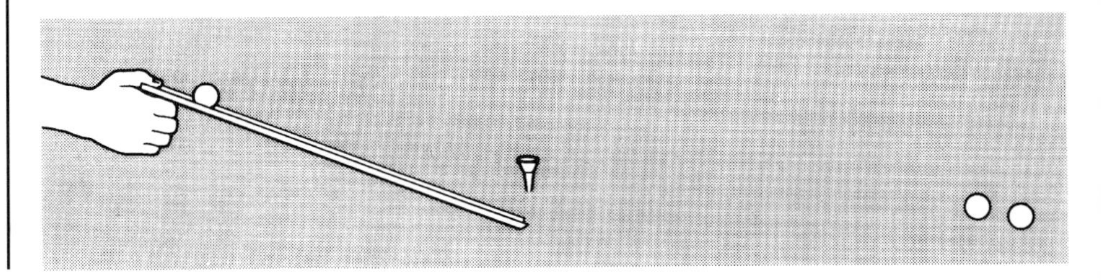

A stimpmeter is a device for measuring the *speed* of a golf green. A ball is allowed to roll down a slope of fixed length (30 inches) and angle to the horizontal (20°), and onto a horizontal green. According to how far the ball travels, the green is classified on a scale between *slow* and *fast.*

Construct a similar device of your own and use it to measure the average force of resistance on a golf ball travelling over a green (or carpet).

By varying the point of release of the ball, investigate whether the resistance to the golf ball may reasonably be modelled as a constant force.

Experiment

Energy Losses

Set up a track like this one.

Release cars or trolleys from different heights and record the heights that they reach on the opposite side. Use your results to formulate a model for the force of resistance acting on them.

KEY POINTS

- The work done by a force F is given by Fs where s is the distance moved in the direction of the force.
- The kinetic energy (K.E.) of a body of mass m moving with speed v is given by $\frac{1}{2}mv^2$. Kinetic energy is the energy a body possesses on account of its motion.
- The work-energy principle states that the total work done by all the forces acting on a body is equal to the increase in the kinetic energy of the body.
- Potential energy is the energy a body possesses on account of its position.
- The gravitational potential energy of a body of mass m at height h above a given reference level is given by mgh.
- Power is the rate of doing work, and is given by Fv.
- The S.I. unit for energy is the joule and that for power is the watt.

5 Impulse and momentum

I collided with a stationary truck coming the other way.

Statement of an insurance form reported in the Toronto News.

The karate expert in the picture has just broken a pile of six roof-tiles with a single blow from his bare hand. Forces in excess of 3000 N have been measured during karate chops. How is this possible?

Impulse

Although the karate expert produces a very large force, it acts for only a short time. This is often the case in situations where impacts occur, as in the following example involving a tennis player.

EXAMPLE

A tennis player hits the ball as it is travelling towards her at 10 ms^{-1} horizontally. Immediately after she hits it, the ball is travelling away from her at 20 ms^{-1} horizontally. The mass of the ball is 0.06 kg. What force does she exert on the ball?

Solution

As it stands, you do not have enough information to answer this question because you do not know for how long the ball is in contact with the racquet.

$u = -10$

$v = 20$

If you estimate that the contact time, t, is $\frac{1}{10}$ of a second, and the force, F, is uniform throughout it, then you can apply the constant acceleration equation $v = u + at$, and Newton's second law.

$$a = \frac{v - u}{t}$$

$$= \frac{20 - (-10)}{\frac{1}{10}}$$

$$= 300$$

Substituting for a in $F = ma$ gives

$$F = 0.06 \times 300 = 18.$$

The magnitude of the constant force would in this case be 18 newtons.

The trouble is that if you estimate a shorter contact time, say $\frac{1}{50}$ of a second (a stab shot), the same procedure gives $a = 1500$ and $F = 90$, a very different answer.

How long is the ball actually in contact with the racquet? The answer is that you cannot know with any degree of accuracy.

Since you cannot know t, you cannot know F either, but what you do know is the effect of F and t together. It is to change a velocity of -10 ms^{-1} into one of $+20$ ms^{-1}.

So what you can say is

$$F = ma \Rightarrow a = \frac{F}{m}$$

Substituting this in

$$v = u + at \text{ gives } v = u + \frac{F}{m}t.$$

Rearranging this,

$$Ft = mv - mu$$

$$= 0.06 \times 20 - 0.06 \times (-10) \quad (1)$$

$$= 1.8.$$

This quantity, force $\times$ time, Ft is called the *impulse* that has acted on the ball. Its unit is the newton second.

The calculations of F for different values of t are somewhat unrealistic because in practice F will not be constant. It will increase as the ball embeds itself in the strings and then decrease as the ball is catapulted away. That however does not alter the conclusion that the ball has received an impulse of 1.8 Ns in the given direction. The fact that you do not know exactly what went on when the ball was in contact with the racquet does not matter, as the impulse was calculated from the outcome of the impact, not from an analysis of the impact itself.

This is typical of many situations where a large force acts for a very short time. Such forces are sometimes called impulsive forces.

The use of the term impulse is not however restricted to forces which act for short time intervals. An impulse may be the effect of a force over any length of time, seconds, minutes, even years.

Impulse is a vector quantity, in the direction of the force. It is often denoted by **J**, with its magnitude written in italics, J.

Momentum

The two quantities on the right hand side of equation (1), mv and mu, are the final momentum and the initial momentum of the body.

The momentum of a body is defined to be the product of its mass and its velocity. Momentum is often thought of as a measure of the quantity of motion that a body carries with it. In the absence of any external force, the momentum of a body is constant. To change the momentum requires a force.

Since velocity is a vector quantity, so also is momentum. The direction of a body's momentum is the same as that of its velocity.

The S.I. unit for momentum is the newton second, the same as that for impulse. 1 Ns is the momentum of a body of mass 1 kg travelling at 1 ms^{-1}.

For Discussion

The magnitude of a body's momentum may be thought of as its resistance to being stopped. A cricket ball of mass 0.15 kg, bowled very fast at 40 ms^{-1}, has momentum 6 Ns, much less than a 20 tonne railway truck moving at the very slow speed of 1 centimetre per second which has momentum 200 Ns.

Which would you rather meet on a dark night, a body with high momentum and low energy, or one with low momentum and high energy?

For Discussion

Explain why a rocket can have constant momentum even though its speed is increasing. What can you say about the velocity of the rocket?

The impulse – momentum equation

As you have seen in the example of the tennis player, the change in momentum is equal to the impulse of the force.

$$\text{impulse} = \text{change in momentum}$$

This is called the impulse–momentum equation.

Impulse can thus be thought of in two ways which are entirely equivalent:

$$\text{impulse} = \text{force} \times \text{time}$$

$$\text{impulse} = \text{change in momentum}$$

For Discussion

Compare the impulse momentum equation with the work-energy principle.

EXAMPLE

A ball of mass 50 grams hits the ground with a speed of 4 ms^{-1} and rebounds with an initial speed of 3 ms^{-1}. If the ball is in contact with the ground for 0.1 seconds,

(i) find the average force exerted on the ball,

(ii) find the loss of kinetic energy during the impact.

Solution

(i) The impulse is given by:

$$J = mv - mu$$

$$= 0.05 \times 3 - 0.05 \times (-4)$$

$$= 0.35$$

3 ms^{-1} + 4 ms^{-1} J Ns

The impulse J is also given by

$$J = Ft$$

where F is the average force, i.e. the constant force which, acting for the same time interval, would have the same effect as the variable force which actually acted.

$$\therefore \quad 0.35 = F \times 0.1$$

$$F = 3.5$$

So the ground exerts an average upward force of 3.5 N.

(ii)

$$\text{Initial K.E.} = \tfrac{1}{2} \times 0.05 \times 4^2 = 0.400 \text{ joules}$$

$$\text{Final K.E.} = \tfrac{1}{2} \times 0.05 \times 3^2 = 0.225 \text{ joules}$$

$$\text{Loss in K.E.} = 0.175 \text{ joules}$$

(This is converted into heat and sound.)

NOTE

This example demonstrates the important point that mechanical energy is not conserved during an impact.

EXAMPLE

A car of mass 800 kg is pushed with a constant force of magnitude 200 Newtons for 10 s. If the car starts from rest, find its speed at the end of the ten second interval.

Solution

Since the force of 200 newtons act for 10 seconds then the impulse on the car is

$$J = 200 \times 10 = 2000 \text{ (in newton seconds).}$$

(The impulse is in the direction of the force).

Hence the change in momentum in newton seconds is

$$mv = 2000$$

$$\therefore \quad v = \frac{2000}{800} = 2.5$$

The speed at the end of the time interval is 2.5 ms^{-1}.

NOTE

This problem could have been solved using Newton's second law. Since $F = ma$ we have $a = 200/800 = 0.25\ ms^{-2}$. Since the acceleration is constant, the final speed is given by $v = u + at$. So with $u = 0$ and $t = 10$ we have $v = 10 \times 0.25 = 2.5\ ms^{-1}$.

EXAMPLE

In a game of snooker the cue ball (W) of mass 0.2 kg is hit towards a stationary red ball (R) at 0.8 ms^{-1}. After the collision the cue ball is moving at 0.6 ms^{-1} having been deflected through 30°.

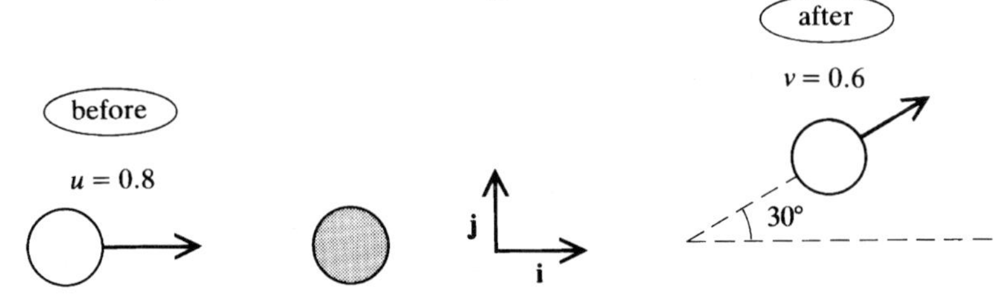

Find the change in momentum of the cue ball and show this change in a vector diagram.

Solution

In terms of unit vectors *i* and *j* the velocities before and after the collision are given by

$$\mathbf{u} = 0.8\mathbf{i}$$
$$\mathbf{v} = 0.6\cos 30°\mathbf{i} + 0.6\cos 60°\mathbf{j}$$

Then $\quad$ impulse = change in momentum

$$= m\mathbf{v} - m\mathbf{u}$$

$$= 0.2(0.6\cos 30°\mathbf{i} + 0.6\cos 60°\mathbf{j}) - 0.2(0.8\mathbf{i})$$

$$= -0.056\mathbf{i} + 0.06\mathbf{j}$$

This change has magnitude 0.082 Ns at an angle of 133° to the initial motion of the ball.

This is shown on the vector diagram below. Note that the impulse-momentum equation shows the direction of the impulsive force acting on the cue ball.

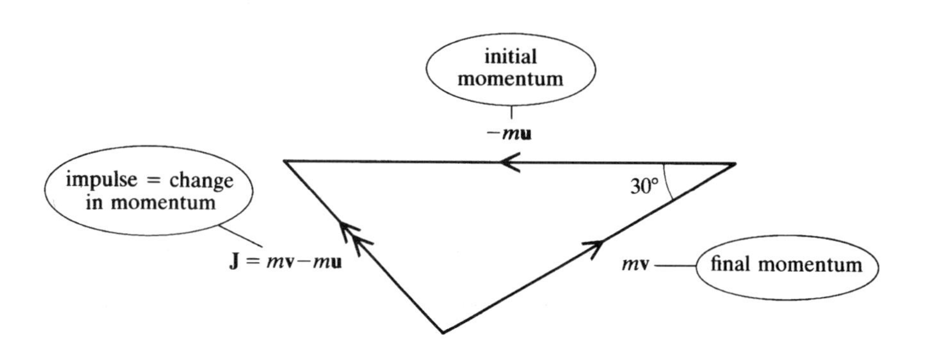

EXAMPLE

A hockey ball of mass 0.15 kg is moving at 4 ms^{-1} parallel to the side of a pitch when it is struck by a blow from a hockey stick that exerts an impulse of 4 Ns at an angle of 120° to its direction of motion. Find the final velocity of the ball.

Solution

The vector diagram shows the motion of the ball.

$u = 4(ms^{-1})$ $\quad$ $J = 4(Ns)$ 120° $\quad$ v ϕ

In terms of unit vectors **i** and **j** we have

$$\mathbf{u} = 4\mathbf{i}$$

$$\text{and } \mathbf{J} = 4\cos 60°\mathbf{i} + 4\cos 30°\mathbf{j}$$

$$= -2\mathbf{i} + 3.46\mathbf{j}$$

Using $J = m\mathbf{v} - m\mathbf{u}$

$$-2\mathbf{i} + 3.464\mathbf{j} = 0.15\mathbf{v} - 0.15 \times 4\mathbf{i}$$

$$\Rightarrow \quad 0.15\mathbf{v} = -2\mathbf{i} + 0.6\mathbf{i} + 3.464\mathbf{j}$$

$$\Rightarrow \quad 0.15\mathbf{v} = -1.4\mathbf{i} + 3.464\mathbf{j}$$

$$\Rightarrow \quad \mathbf{v} = -9.33\mathbf{i} + 23.1\mathbf{j}$$

The magnitude of the velocity is given by $v = \sqrt{9.33^2 + 23.1^2}$

$= 24.9\ (\text{ms}^{-1})$.

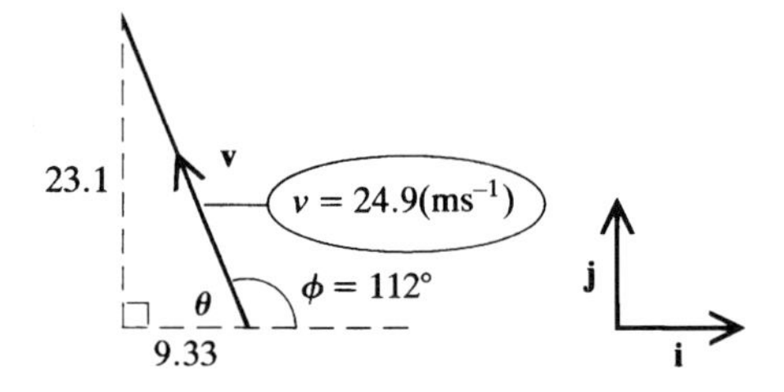

The angle, θ, is given by $\theta = \tan^{-1}\left(\dfrac{23.1}{9.33}\right)$

$= 68°$

$\phi = 180° - 68°$

$= 112°$.

After the blow, the ball has a velocity of magnitude 24.9 ms^{-1} at an angle of 112° to the original direction of motion.

Variable forces

Very often the forces that change the motion of objects are not constant. When a trampolinist hits the bed of a trampoline, the upward force acting on her is linked to the deformation of the bed and the extension of the springs.

The greater the deformation and extension, the greater the force. While it is possible to deal with situations like this by considering the average force, it is also possible to deal precisely with variable forces.

The motion of a particle under a variable force

Motion under a variable is also covered by the impulse – momentum equation. Think of a particle of mass m subject to a variable force F for a time T.

Suppose that during this time the velocity of the particle changes from U to V.

Applying Newton's Second Law

$$F = ma = m\frac{dv}{dt}$$

Integrating between $t = 0$ and $t = T$ we have

$$\int_0^T F dt = \int_0^T m\frac{dv}{dt}dt$$

$$= \int_{v=U}^{v=V} m dv$$

$$= mV - mU$$

Since when $t = 0$, velocity = U and when $t = T$, velocity = V.

The right hand side of the equation is the change in momentum and the left hand side, $\int_0^T F dt$, is the impulse. The impulse is therefore the area under the graph against F.

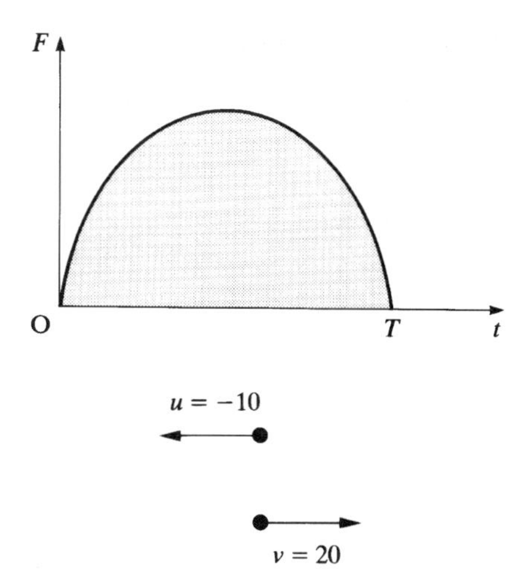

Figure 5.1

Note that if the force, F, is constant then the impulse

$$= \int_0^T F\mathrm{d}t = F\int_0^T \mathrm{d}t = FT \quad \text{as before.}$$

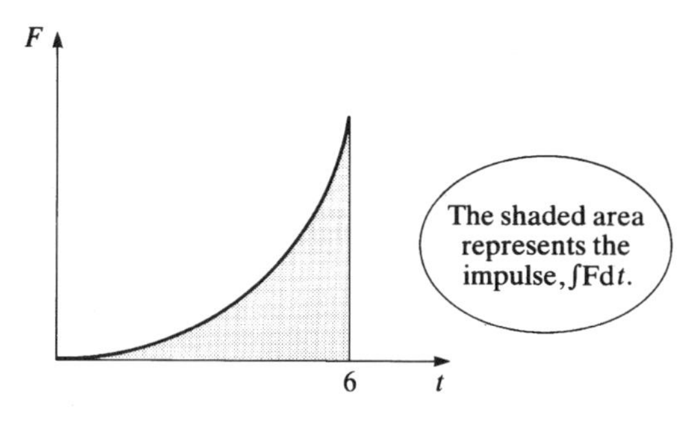

Exercise 5A

1. Find the linear momentum of the following objects, assuming each of them to be travelling in a straight line.
(i) An ice skater of mass 50 kg travelling with speed 10 ms^{-1}.
(ii) An elephant of mass 5 tonnes moving at 4 ms^{-1}.
(iii) A train of mass 7000 tonnes travelling at 40 ms^{-1}.
(iv) A bacterium of mass 2×10^{-16} g moving with speed 1 mms^{-1}.

2. Calculate the impulse required in each of these situations:
(i) to stop a car of mass 1.3 tonnes travelling at 14 ms^{-1};
(ii) to putt a golf ball of mass 1.5 g with speed 1.5 ms^{-1};
(iii) to stop a cricket ball of mass 0.15 kg travelling at 20 ms^{-1};
(iv) to fire a bullet of mass 25 g with speed 400 ms^{-1}.

3. A stone of mass 1.5 kg is dropped from rest. After a time interval t s, it has fallen a distance s m and has velocity v ms^{-1}. Taking g to be 10 ms^{-2} and neglecting air resistance:
(i) write down the force F in newtons acting on the stone;
(ii) find the distance, s, that the stone has fallen when $t = 2$;
(iii) find the velocity, v in ms^{-1}, of the stone when $t = 2$;
(iv) write down the value, units and meaning of Fs and explain why this has the same value as $\frac{1}{2} \times 1.5v^2$.
(v) write down the value, units and meaning of Ft and explain why this has the same value as $1.5v$.

4. A girls throws a ball of mass 0.06 kg vertically upwards with initial speed 20 ms^{-1}.

Take g to be 10 ms^{-2} and neglect air resistance.

(i) What is the initial momentum of the ball?
(ii) How long does it take for the ball to reach the top of its flight?
(iii) What is the momentum of the ball when it is at the top of its flight?
(iv) What impulse acted on the ball over the period between its being thrown and its reaching maximum height?
(v) What is the average force acting on the ball during this period? How would you usually describe it?

5. A netball of mass 425 g is moving horizontally with speed 5 ms^{-1} when it is caught.
(i) Find the impulse needed to stop the ball.
(ii) Find the average force needed to stop the ball if it takes
(a) 0.1 s (b) 0.05 s.
(iii) Why does the action of taking a ball into your body make it easier to catch?

6. A car of mass 0.9 tonnes is travelling at 13.2 ms^{-1} when it crashes head on into a wall. The car is brought to rest in a time of 0.12 s. Find
(i) the impulse acting on the car;
(ii) the average force acting on the car;
(iii) the average deceleration of the car in terms of g, (taken to be 10 ms^{-2}).
(iv) Explain why many cars are designed with crumple zones rather than with completely rigid construction.

7. Boris is sleeping on a bunk bed at a height of 1.5 m when he rolls over and falls out this mass is 20 kg.
(i) Find the speed with which he hits the floor.
(ii) Find the impulse that the floor has exerted on him when he has come to rest.
(iii) Find the impulse he has exerted on the floor.

It takes Boris 0.2 seconds to come to rest.
(iv) Find the average force acting on him during this time.

8. A railway truck of mass 10 tonnes is travelling at 3 ms^{-1} along a siding when it hits some buffers. After the impact it is travelling at 1.5 ms^{-1} in the opposite direction.
(i) Find the initial momentum of the truck, remembering to specify its direction.
(ii) Find the momentum of the truck after it has left the buffers.
(iii) Find the impulse that has acted on the truck.

During the impact the force F N that the buffers exert on the truck varies as shown in this graph.

Exercise 5A continued

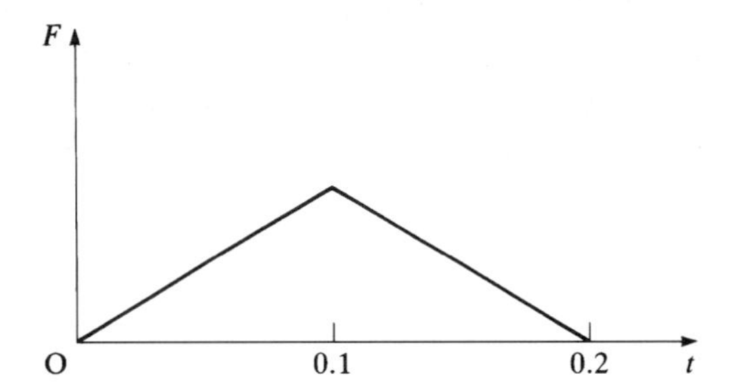

(iv) State what information is given by the area under the graph.
(v) What is the greatest value of the force F?

9. A snooker ball of mass 0.08 kg is travelling with speed $3.5\ \mathrm{ms}^{-1}$ when it hits the cushion at an angle of $60°$. After the impact the ball is travelling with speed $2\ \mathrm{ms}^{-1}$ at an angle $30°$ to the cushion.

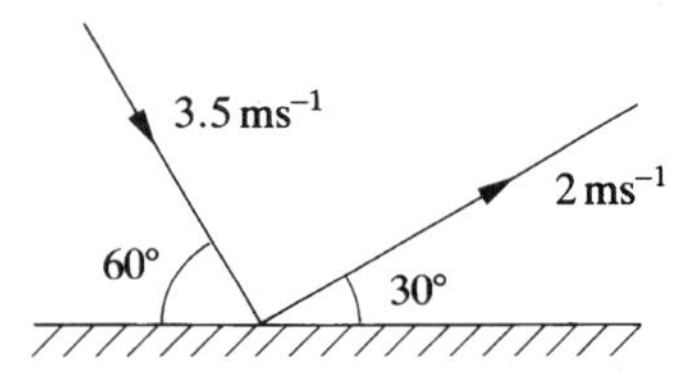

(i) Draw accurate scale diagrams to represent the following vectors:
  (a) the momentum of the ball before impact;
  (b) the momentum of the ball after impact;
  (c) the change in momentum of the ball during impact.
(ii) Use your answer to part (i) (c) to *estimate* the magnitude and direction of the impulse acting on the ball.
(iii) Resolve the velocity of the ball before and after impact into components parallel and perpendicular to the cushion.
(iv) Use your answers to part (iii) to *calculate* the impulse which acts on the ball during its impact with the cushion. Comment on your answers.

10. A hockey ball of mass 0.15 kg is travelling with velocity $12\mathbf{i} - 8\mathbf{j}$ in metres per second, where the unit vectors $\mathbf{i}$ and $\mathbf{j}$ are in horizontal directions parallel and perpendicular to the length of the pitch, and the vector $\mathbf{k}$ is vertically upwards. The ball is hit by Jane with an impulse $-4.8\mathbf{i} + 1.2\mathbf{j}$.
(i) What is the velocity of the ball immediately after Jane has hit it?

The ball goes straight, without losing any speed, to Fatima in the opposite team who hits it without stopping it. Its velocity is now $14\mathbf{i} + 4\mathbf{j} + 3\mathbf{k}$.
(ii) What impulse does Fatima give the ball?
(iii) Which player hits the ball harder?

11. A hailstone of mass $4\,\mathrm{g}$ is travelling with speed $20\,\mathrm{ms^{-1}}$ when it hits a window as shown in the diagram. It bounces off the window; the vertical component of the velocity is unaltered, but the horizontal component is now $2\,\mathrm{ms^{-1}}$ away from the window.

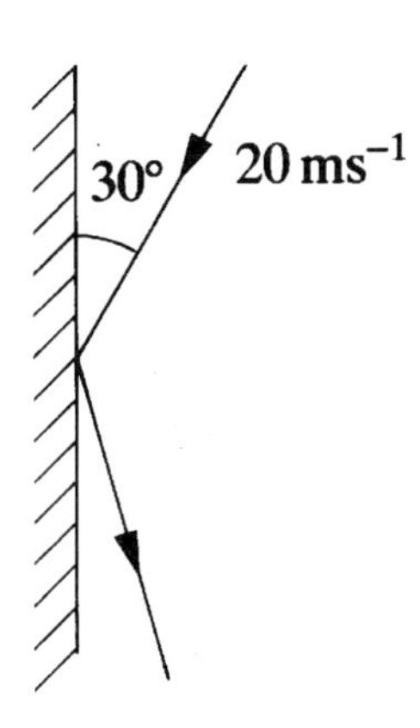

(i) State the magnitude and direction of the impulse

(a) of the window on the hailstone,

(b) of the hailstone on the window.

At the peak of the storm, hailstones like this are hitting the window at the rate of 540 per minute.

(ii) Find the average force of the hail on the window.

12. The graph shows the magnitude of the driving force on a van during the first 4 seconds after it starts from rest. The mass of the van is $2500\,\mathrm{kg}$.

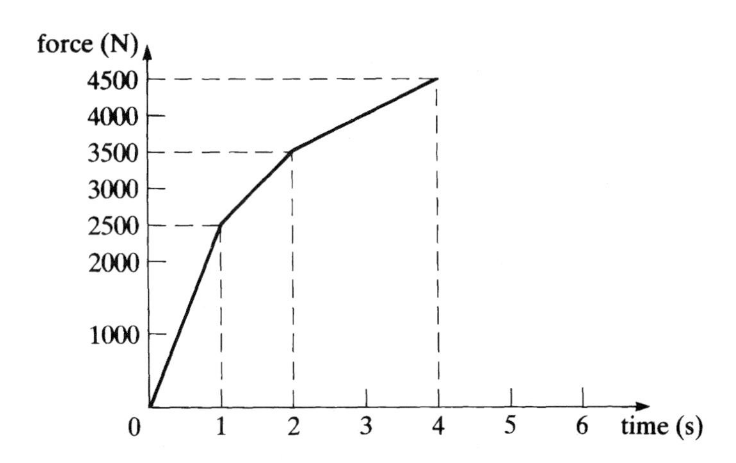

(i) What information is given by the area under the graph?

(ii) Find the total impulse on the van over the whole interval.

(iii) Find the final speed of the van, ignoring the effect of air resistance.

Conservation of momentum

Collisions

For Discussion

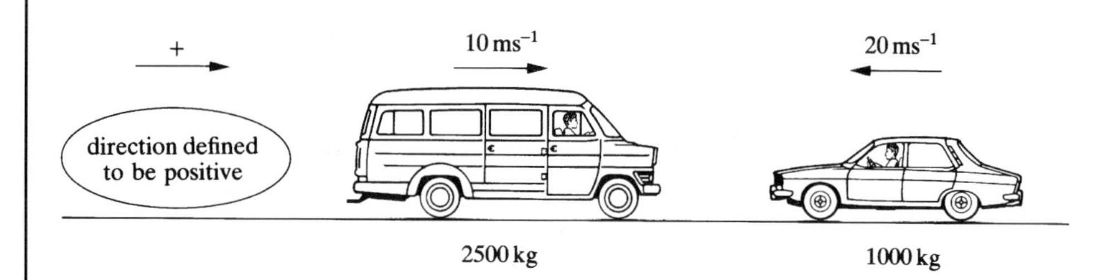

These two vehicles collide head-on. How would you investigate this situation? Can you find a relationship between the change in momentum of the van and that of the car?

Commentary

The first thing to remember is Newton's third law. The force that body A exerts on body B is equal to the force that B exerts on A, but in the opposite direction.

Suppose that once the van is in contact with the car, it exerts a force F on the car for a time t. Newton's third law tells us that the car also exerts a force F on the van for a time t. (This applies whether F is constant or variable). So both vehicles receive equal impulses, but in opposite directions. Consequently the increase in momentum of the car in the positive direction is exactly equal to the increase in momentum of the van in the negative direction. For the two vehicles together, the total change in momentum is zero.

This example illustrates the law of conservation of momentum, which states that when there are no external influences on a system, the total momentum of the system is constant. Since momentum is a vector quantity, this applies to the magnitude of the momentum in any direction.

For a collision, you can say

$$\text{total momentum of system before collision} = \text{total momentum of system after collision}$$

It is important to remember, however, that although momentum is conserved in a collision, mechanical energy is not conserved.

EXAMPLE

The two vehicles in the previous discussion collide head-on, and as a result the van comes to rest.

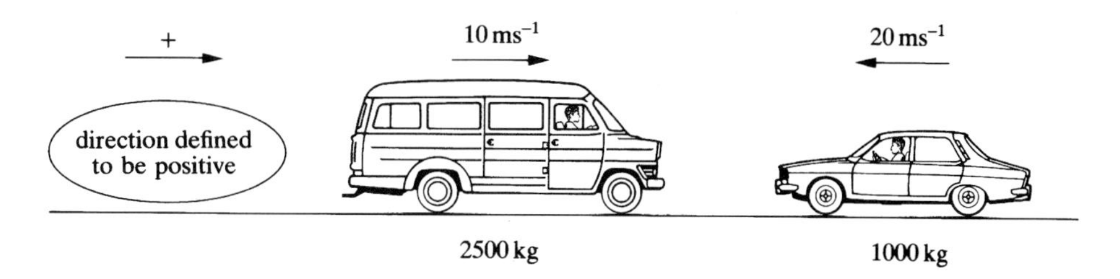

Find

(i) the final velocity of the car, v ms^{-1};

(ii) the impulse on each vehicle;

(iii) the kinetic energy lost.

If it is assumed that the impact lasts for one twentieth of a second, find

(iv) the force on each vehicle and its acceleration.

Solution

(i)

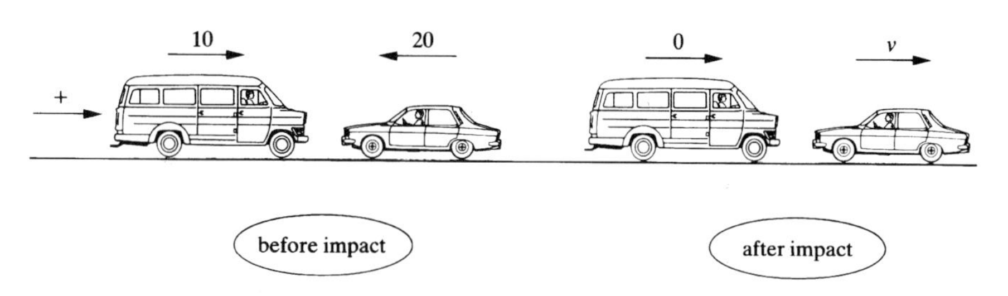

Using conservation of momentum, and taking the positive direction as being to the right:

$$\begin{aligned} 2500 \times 10 + 1000 \times (-20) &= 2500 \times 0 + 1000 \times v \\ 5000 &= 1000v \\ v &= 5 \end{aligned}$$

The final velocity of the car is 5 ms^{-1} in the positive direction (ie. the car travels backwards).

(ii)

$$\begin{aligned} \text{Impulse} &= \text{change in momentum} \\ &= \text{final momentum} - \text{initial momentum} \\ \text{For the van, impulse} &= 2500 \times 0 - 2500 \times 10 \\ &= -25\,000 \text{ Ns.} \\ \text{For the car, impulse} &= 1000 \times 5 - 1000 \times (-20) \\ &= +\,25\,000 \text{ Ns.} \end{aligned}$$

The van experiences an impulse of 25 000 Ns in the negative direction, the car an equal and opposite impulse.

(iii)

$$\text{Initial K. E.} = \frac{1}{2} \times 2500 \times 10^2 + \frac{1}{2} \times 1000 \times 20^2$$
$$= 325\,000 \text{ J}$$

$$\text{Final K. E.} = \frac{1}{2} \times 2500 \times 0^2 + \frac{1}{2} \times 1000 \times 5^2$$
$$= 12\,500 \text{ J}$$
$$\text{Loss in K. E.} = 312\,500 \text{ J}$$

(iv)

$$\text{Impulse} = \text{average force} \times \text{time}$$
$$25\,000 = F \times \frac{1}{20}$$
$$F = 500\,000 \text{ N}$$

(acting to the right on the car and to the left on the van).

Using $F = ma$ on each vehicle gives an acceleration of 500 ms^{-2} for the car and -200 ms^{-2} for the van.

For Discussion

These accelerations (500 ms^{-2} and -200 ms^{-2}) seem very high. Are they realistic for a head-on collision?

Work out the distance each car travels during the time interval of one twentieth of a second between impact and separation. This will give you an idea of the amount of damage there would be.

Is it better for cars to be made strong so that there is little damage, or to be designed to buckle under impact?

Experiment

Set up the apparatus as shown below. Truck A can be released from the top of the slope so that it always hits truck B at the same speed.

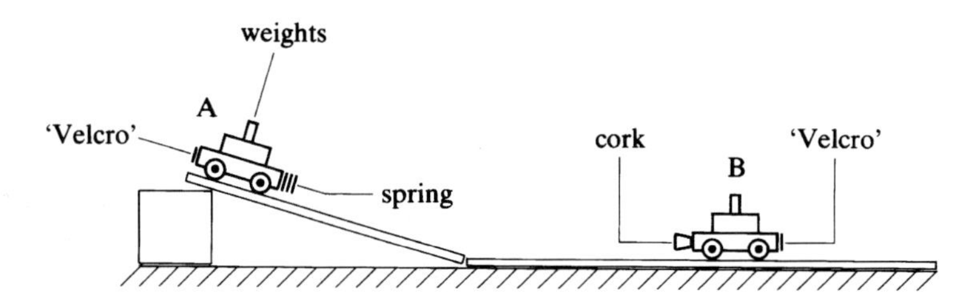

For each situation below describe what you think will happen and then test your prediction.

1. Arrange the trucks so that the spring on truck A hits the cork on truck B.

(i) Load both trucks with the same mass, then release A so that it rolls down and hits B.

(ii) Now load B so that it is very much heavier than A.
(iii) Now load A so that it is very much heavier than B.

2. Arrange the trucks so that the velcro tabs hit each other during the collision. Now repeat experiments (i) to (iii) above.

3. Rearrange the track as shown below.

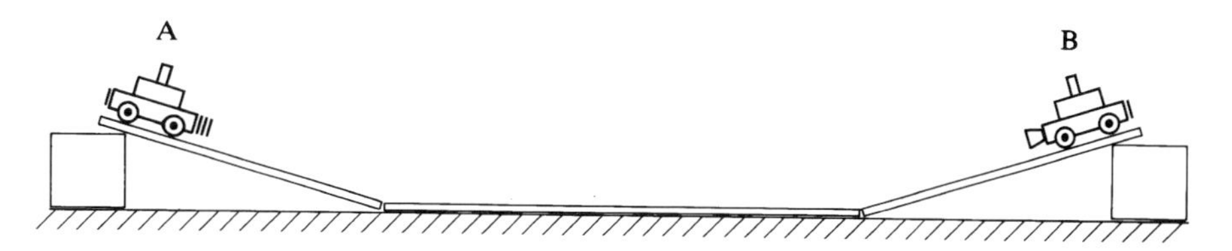

(i) Release trucks that are loaded equally from both ends of the track, so that the spring and cork come into contact.
(ii) Repeat (i), but with one truck loaded so that it is much heavier than the other.
(iii) Release equally loaded trucks from each end so that the velcro ends come into contact.
(iv) Repeat (iii), but with one truck loaded so that it is much heavier than the other.

EXAMPLE

In an experiment on lorry bumper design, the Transport Research Laboratory arranged for a car and a lorry, of masses 1 and 3.5 tonnes to travel towards each other, both with speed $9\ \text{ms}^{-1}$. After a head-on collision both vehicles move together at approximately $5\ \text{ms}^{-1}$ in the direction that the lorry was originally moving. Show that the total momentum is conserved during the collision.

Solution

The situation before the collision is illustrated below.

Taking the positive direction to be to the right, the momentum of the car in Ns is

$$1000 \times 9 = 9000$$

The momentum of the lorry in Ns is

$$3500 \times (-9) = -31\,500$$

The total momentum of the system in Ns is therefore:

$$9000 - 31\,500 = -22\,500$$

After the collision the situation can be represented as a single object of mass 4.5 tonnes moving at -5 ms^{-1}.

The momentum in Ns of the system after the collision is

$$4500 \times (-5) = -22\,500$$

Clearly these figures are the same, which confirms that momentum is conserved.

EXAMPLE

A child of mass 30 kg running through a supermarket at 4 ms^{-1} leaps onto a stationary shopping trolley of mass 15 kg. Find the speed of the child and trolley together, assuming that the trolley is free to move easily.

Solution

The diagram shows the situation before the child hits the trolley.

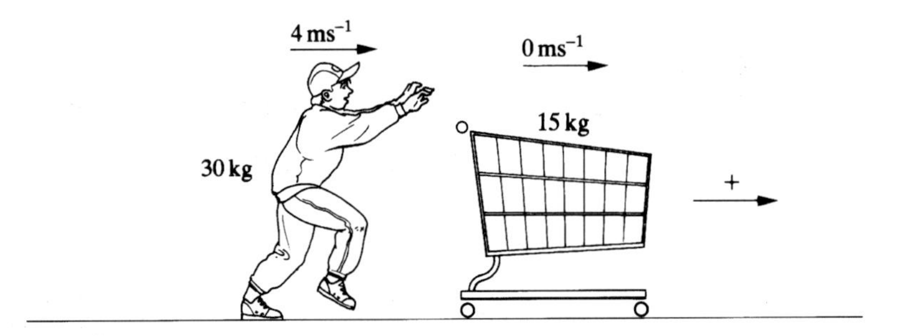

Taking the direction of the child's velocity as positive, total momentum (in Ns) before impact

$$= 4 \times 30 + 0 \times 15$$
$$= 120.$$

The situation after impact is shown below.

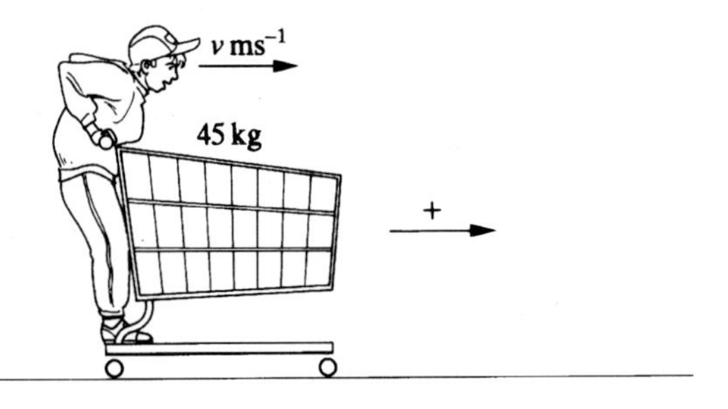

Total mass of child and trolley together = 45 kg.

Total momentum after impact = $45v$.

Conservation of momentum gives:

$$120 = 45v$$
$$v = \frac{120}{45}$$
$$= 2\tfrac{2}{3}.$$

The child and the trolley together move at 2 ⅔ ms^{-1}.

Explosions

Conservation of momentum also applies when explosions take place provided there are no external forces. For example when a bullet is fired from a rifle, or a rocket is launched.

EXAMPLE

A rifle of mass 8 kg is used to fire a bullet of mass 80 g at a speed of 200 ms^{-1}. Calculate the initial recoil speed of the rifle.

Solution

Before the bullet is fired the total momentum of the system is zero.

After the firing the situation is as illustrated below.

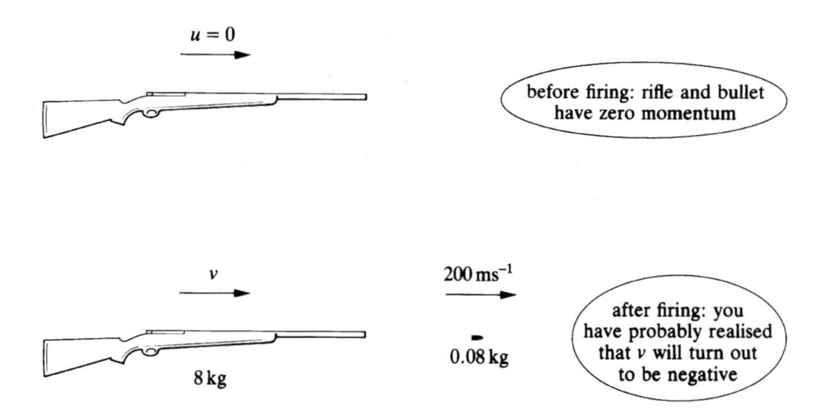

The total momentum in the positive direction after the firing is $0.08 \times 200 + 8v$.

For momentum to be conserved,

$$200 \times 0.08 + 8v = 0$$

so that

$$v = \frac{-200 \times 0.08}{8}$$
$$= -2$$

The final recoil speed of the rifle is 2 ms^{-1}.

Exercise 5B

1. A railway truck of mass 20 tonnes is shunted with speed 3 ms^{-1} towards a stationary truck of mass 10 tonnes. After impact the two trucks remain in contact. What is their speed?

2. The driver of the car of mass 1000 kg falls asleep while it is travelling at 30 ms^{-1}. The car runs into the back of the car in front which has mass 800 kg and is travelling in the same direction at 20 ms^{-1}. The bumpers of the two cars become locked together and they continue as one vehicle.
(i) What is the final speed of the cars?
(ii) What impulse does the larger car give to the smaller one?
(iii) What impulse does the smaller car give to the larger one?

3. A spaceship of mass 50 000 kg travelling with speed 200 ms^{-1} docks with a space station of mass 500 000 kg travelling in the same direction with speed 195 ms^{-1}. What is their speed after the docking is completed.

4. A lorry of mass 5 tonnes is towing a car of mass 1 tonne. Initially the tow rope is slack and the car stationary. As the rope becomes taut the lorry is travelling at 2 ms^{-1}.
(i) Find the speed of the car once it is being towed.
(ii) Find the magnitude of the impulse transmitted by the tow rope and state the direction of the impulse on each vehicle.

5. A bullet of mass 50 g is moving horizontally at 200 ms^{-1} when it becomes embedded in a stationary block of mass 16 kg which is free to slide on a smooth horizontal table.
(i) Calculate the speed of the bullet and the block after the impact.
(ii) Find the impulse from the bullet on the block.

The bullet takes 0.01 s to come to rest relative to the block.
(iii) What is the average force acting on the bullet while it is decelerating?

6. A spaceship of mass 50 000 kg is travelling through space with speed 5000 ms^{-1} when a crew member throws a box of mass 5 kg out of the back with speed 10 ms^{-1} relative to the spaceship.
(i) What is the absolute speed of the box?
(ii) What is the speed of the spaceship after the box has been thrown out?

7. A gun of mass 500 kg fires a shell of mass 5 kg horizontally with muzzle speed 300 ms^{-1}.
(i) Calculate the recoil speed of the gun.

An army commander would like his soldiers to be able to fire such a shell from a rifle held against their shoulders (to enable them to attack armoured vehicles).

(ii) Explain why such an idea has no hope of success.

8. Manoj (mass 70 kg) and Alka (mass 50 kg) are standing stationary facing each other on a smooth ice rink. They then push against each other with a force of 35 N for 1.5 s. The direction in which Manoj faces is taken as positive.

(i) What is their total momentum before they start pushing?

(ii) Find the velocity of each of them after they have finished pushing.

(iii) Find the momentum of each of them after they have finished pushing.

(iv) What is their total momentum after they have finished pushing?

9. Katherine (mass 40 kg) and Elisabeth (mass 30 kg) are on a sledge (mass 10 kg) which is travelling across smooth horizontal ice at 5 ms^{-1}. Katherine jumps off the back of the sledge with speed 4 ms^{-1} backwards relative to the sledge.

(i) What is Katherine's absolute speed when she jumps off?

(ii) With what speed does Elisabeth, still on the sledge, then go?

Elisabeth then jumps off in the same manner, also with speed 4 ms^{-1} relative to the sledge.

(iv) What is the speed of the sledge now?

(v) What would the final speed of the sledge have been if Katherine and Elisabeth had both jumped off at the same time, with speed 4 ms^{-1} backwards relative to the sledge?

10. A pile-driver has a block of mass 2 tonnes which is dropped from a height of 5 m onto the pile of mass 600 kg which it is driving vertically into the ground. The block rebounds with a speed of 2 ms^{-1} immediately after the impact. Taking g to be 10 ms^{-2} find:

(i) find the speed of the block immediately before the impact;

(ii) find the impulse acting on the block;

(iii) find the impulse acting on the pile.

From the moment of impact the pile takes 0.025 s to come to rest.

(iv) Calculate the force of resistance on the pile, assuming it to be constant.

(v) How far does the pile move?

11. The diagram shows Nicholas, whose mass is 80 kg, standing at the front of a sleigh of length 5 m and mass 40 kg. The sleigh is initially stationary and on smooth ice. Nicholas then walks towards the back of the sleigh with speed 1 ms^{-1} relative to the sleigh.

Exercise 5B continued

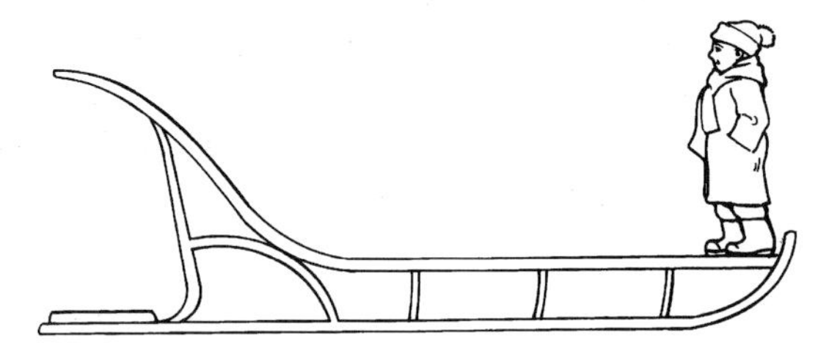

(i) Find the velocity of the sleigh while Nicholas is walking towards the back of it.
(ii) Show that, throughout his walk, the combined centre of mass of Nicholas and the sleigh does not move.
(iii) Investigate whether the result in part (ii) is true in general for this type of situation, or is just a fluke depending on the particular values given to the variables involved.

12. A truck P of mass 2000 kg starts from rest and moves down an incline from A to B as illustrated in the diagram. The distance from A to B is 50 m and $\sin\alpha = 0.05$. CBDE is horizontal.

Neglecting resistance to motion, calculate
(i) the potential energy lost by the truck P as it moves from A to B;
(ii) the speed of the truck P at B.

Truck P then continues from B without loss of speed towards a second truck Q of mass 1500 kg at rest at D. The two trucks collide and move on towards E together. Still neglecting resistances to motion, calculate
(i) the common speed of the two trucks just after they become coupled together;
(ii) the percentage loss of kinetic energy in the collision. [MEI]

Newton's Law of Impact

For Discussion

If you drop two different balls, say a tennis ball and a cricket ball, from the same height, will they both rebound to the same height? How will the heights of the second bounces compare with the heights of the first ones?

Your own experience probably tells you that different balls will rebound to different heights. For example, a tennis ball will rebound to a greater height than a cricket ball. Furthermore, the surface on which the ball is dropped will affect the bounce. A tennis ball dropped onto a concrete floor will rebound higher than if dropped onto a carpeted floor. The following experiment allows you to look at this situation more closely.

Experiment

The aim of this experiment is to investigate what happens when balls bounce.

1. Drop a ball from a variety of heights and record the heights of release h_i and rebound h_r in a table. Repeat several times for each height.
2. Use your values of h_i and h_r, to calculate w_i and w_r, the speeds on impact and rebound. Enter the results in your table.

$$v^2 = u^2 + 2as$$

3. Calculate the ratio w_r / w_i for each pair of headings of h_i and h_r and enter the results in your table.
4. Repeat the experiment with different types of ball.
5. What do you notice about the ratios w_r / w_i ?

Coefficient of Restitution

Newton's experiments on collisions led him to formulate a simple law relating to the velocites before and after a direct collision between 2 bodies, called *Newton's law of impact.*

$$\frac{\text{speed of separation}}{\text{speed of approach}} = \text{constant}$$

This constant is called the *coefficient of restitution* and is conventionally denoted by the letter e. For two particular surfaces, e is a constant between 0 and 1. It does not have units, being the ratio of two speeds.

For very bouncy balls, e is close to 1, and for balls that do not bounce, e is close to 0. A collision for which $e = 1$ is called perfectly elastic, and a collision for which $e = 0$ is called perfectly inelastic.

The value of e for the ball that you used in the experiment is given by w_r/w_i, and you should have found that this had approximately the same value for each time for any particular ball.

NOTE

Newton's law of impact can be written in vector form. If a ball rebounds from a stationary surface in the **j** *direction as shown in figure 5.2, where*

$$\mathbf{v} = v_1\,\mathbf{i} + v_2\,\mathbf{j} \text{ and } \mathbf{u} = u_1\,\mathbf{i} + u_2\,\mathbf{j}\text{, then}$$

$$v_1\mathbf{i} + v_2\mathbf{j} = -eu_1\mathbf{i} + u_2\mathbf{j}$$

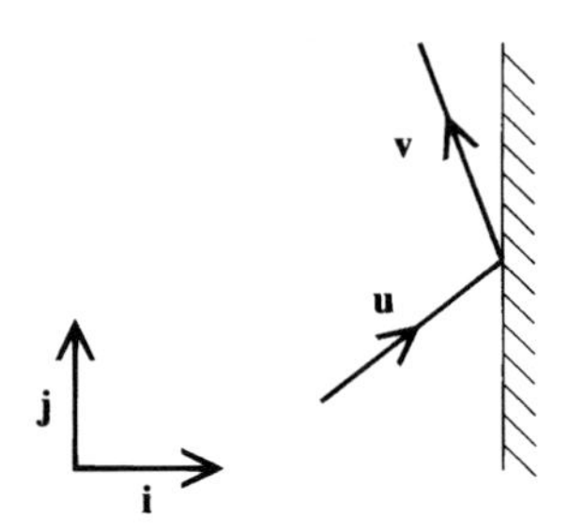

Figure 5.2

Notice that there is no change in velocity parallel to the surface. Newton's law applies to the velocity perpendicular to the surface.

Collisions between two bodies moving in the same straight line

The diagram shows two objects that collide while moving along a straight line. Object A is catching up with B, and after the collision either B moves away from A or they continue together.

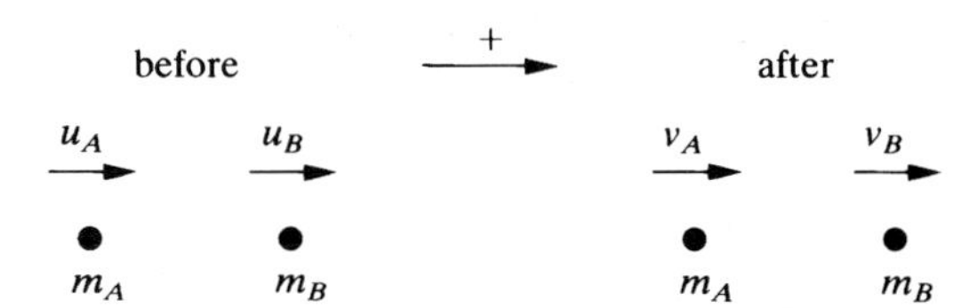

Figure 5.3

Before the collision the speed of approach is $u_A - u_B$.
After the collision the speed of separation is $v_B - v_A$.

Applying Newton's law

$$\frac{\text{speed of separation}}{\text{speed of approach}} = \frac{v_B - v_A}{u_A - u_B} = e$$

$$\Rightarrow \qquad v_B - v_A = e\,(u_A - u_B) \qquad (1)$$

A second equation relating the velocities follows from the law of conservation of momentum in the positive direction (⇒):

$$\text{momentum after collision} = \text{momentum before collision}$$

$$m_A v_A + m_B v_B = m_A u_A + m_B u_B \qquad (2)$$

These two equations, ① and ②, allow you to calculate the final velocities, v_A and v_B, after any collision as shown in the next two examples.

EXAMPLE

A direct collision takes place between two snooker balls. The cue ball travelling at 2ms^{-1} hits a stationary red ball. After the collision the red ball moves in the direction in which the cue ball was moving before the collision. Assume that the balls have equal mass, and that the coefficient of restitution between the two balls is 0.6. Predict the velocities of the two balls after the collision.

Solution

Let the mass of each ball be m, and call the (white) cue ball 'W' and the red ball 'R'.

The situation is summarised in the diagram below.

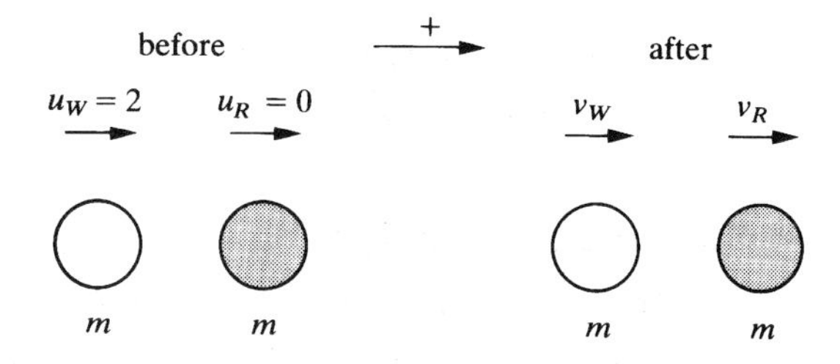

Speed of approach $= 2 - 0 = 2$; speed of separation $= v_R - v_W$

Conservation of momentum gives

$$mv_W + mv_R = mu_W + mu_R$$

Dividing through by m, this becomes

$$v_W + v_R = 2 \qquad ①$$

Newton's law of impacts states that

$$\frac{\text{speed of separation}}{\text{speed of approach}} = e$$

$$\Rightarrow \qquad \Rightarrow \frac{v_R - v_W}{2} = 0.6$$

$$\Rightarrow \qquad \Rightarrow v_R - v_W = 1.2 \qquad ②$$

Adding equations ① and ② gives $2v_R = 3.2$,

so $v_R = 1.6$, and from equation ②, $v_W = 0.4$.

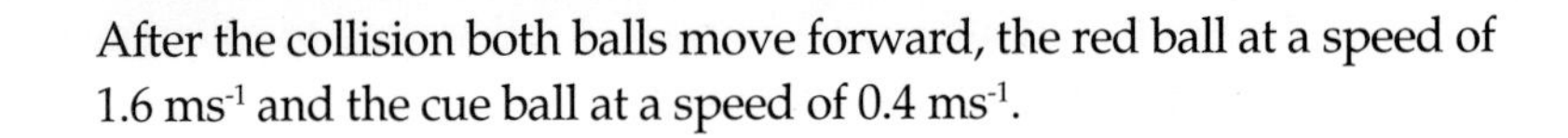
After the collision both balls move forward, the red ball at a speed of 1.6 ms^{-1} and the cue ball at a speed of 0.4 ms^{-1}.

EXAMPLE

An object A of mass m moving with speed $2u$ hits an object B of mass $2m$ moving with speed u in the same direction as A.

(i) Show that object A will continue to move forward whatever the value of e.

(ii) Find the loss of kinetic energy in the case when $e = \frac{1}{2}$.

Solution

(i) Let the velocities of A and B after the collision be v_A and v_B respectively.

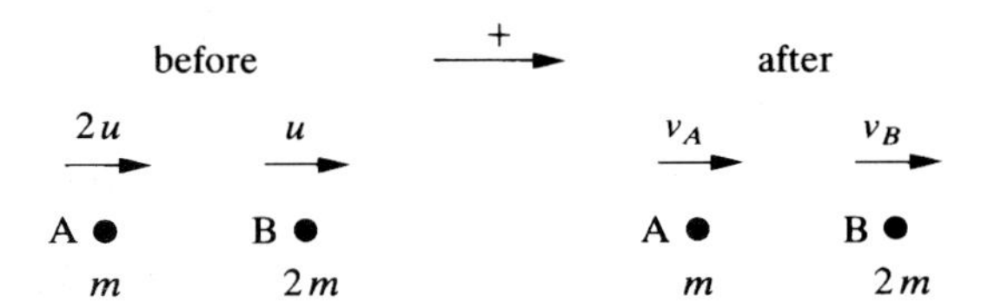

Speed of approach $= 2u - u = u$;
Speed of separation $= v_B - v_A$.

Conservation of momentum gives

$$m(2u) + 2mu = m v_A + 2m v_B$$

Dividing by m gives

$$v_A + 2v_B = 4u \qquad (1)$$

Using Newton's law of impacts

$$\frac{\text{speed of separation}}{\text{speed of approach}} = e$$

$$\frac{v_B - v_A}{u} = e$$

$$\Rightarrow \qquad v_B - v_A = eu \qquad (2)$$

Adding (1) and (2) gives

$$3v_B = (4 + e)u$$

$$v_B = \frac{(4+e)u}{3}$$

From (2),

$$v_A = v_B - eu$$

$$= \frac{(4 + e - 3e)u}{3}$$

$$= \frac{(4 - 2e)u}{3}.$$

Since $0 \le e \le 1, 4 - 2e > 0$ whatever the value of e. So v_A is positive (A continues to move forwards) for all values of e.

(ii) When $e = \frac{1}{2} h$

$$v_A = \frac{(4-1)}{3}u = u$$

$$v_B = \frac{(4+\frac{1}{2})}{3}u = \frac{3u}{2}$$

Initial K. E. of A $= \frac{1}{2} m \times (2u)^2 \quad = 2mu^2$

Initial K. E. of B $= \frac{1}{2} (2m) \times u^2 \quad = mu^2$

Total K. E before impact $= 3mu^2$.

Final K. E. of A $= \frac{1}{2} m \times v_A^2 = \frac{1}{2} mu^2$

Final K. E. of B $= \frac{1}{2} (2m) \times v_B^2 = \frac{9}{4} mu^2$

Total K. E. after impact $= \frac{11}{4} mu^2$

Loss of K. E. $= \left(3 - \frac{11}{4}\right) mu^2 = \frac{1}{4} mu^2$

NOTE

Kinetic energy is lost in any collision in which the coefficient of restitution is not equal to 1.

Exercise 5C

1. In each of the situations shown below, find the unknown quantity, either the initial speed u, the final speed v or the coefficient of restitution e.

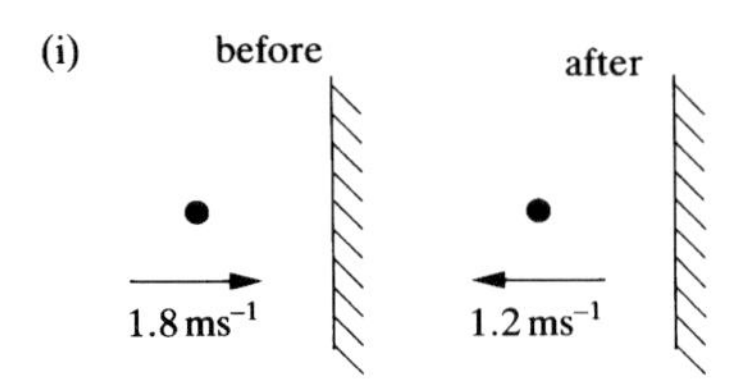

(ii) before after

2.4 ms⁻¹ $v = ?$

$e = 0.6$

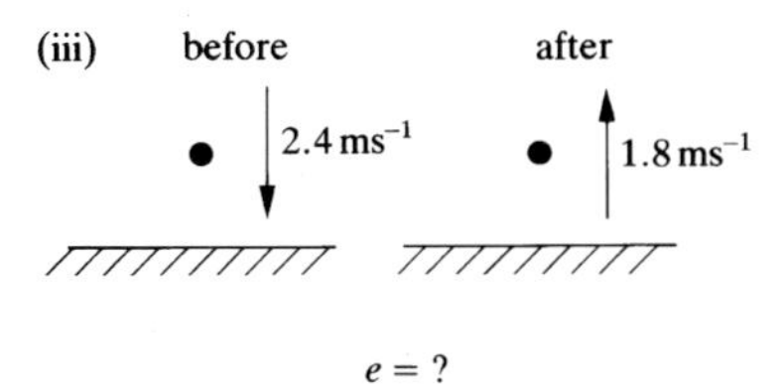

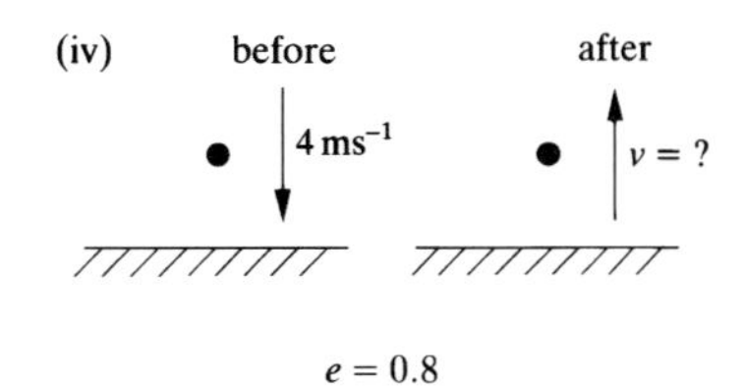

Exercise 5C continued

2. Find the coefficient of restitution in the following situations.
(i) A football hits the goalpost at $10\ \mathrm{ms^{-1}}$ and rebounds in the opposite direction with speed $3\ \mathrm{ms^{-1}}$.
(ii) A beanbag is thrown against the wall with speed $5\ \mathrm{ms^{-1}}$ and falls straight down to the ground.
(iii) A superball is dropped onto the ground, landing with speed $8\ \mathrm{ms^{-1}}$ and rebounding with speed $7.6\ \mathrm{ms^{-1}}$.
(iv) A photon approaches a mirror along a line normal to its surface with speed $3 \times 10^8\ \mathrm{ms^{-1}}$ and leaves it along the same line with speed $3 \times 10^8\ \mathrm{ms^{-1}}$.

3. A tennis ball of mass $60\ \mathrm{g}$ is hit against a practice wall. At the moment of impact it is travelling horizontally with speed $15\ \mathrm{ms^{-1}}$. Just after the impact its speed is $12\ \mathrm{ms^{-1}}$, also horizontally. Find
(i) the coefficient of restitution between the ball and the wall;
(ii) the impulse acting on the ball;
(iii) the loss of kinetic energy during the impact.

4. A ball of mass $80\ \mathrm{g}$ is dropped from a height of $1\ \mathrm{m}$ onto a level floor and bounces back to a height of $0.81\ \mathrm{m}$. Find
(i) the speed of the ball just before it hits the floor;
(ii) the speed of the ball just after it has hit the floor;
(iii) the coefficient of restitution;
(iv) the change in the kinetic energy of the ball from just before it hits the floor to just after it leaves the floor;
(v) the change in the potential energy of the ball from the moment when it was dropped to the moment when it reaches the top of its first bounce;
(vi) the height of the ball's next bounce.

5. Two children drive dodgems straight at each other, and collide head-on. Both dodgems have the same mass (including their drivers) of $150\ \mathrm{kg}$. Isobel is driving at $3\ \mathrm{ms^{-1}}$, Stuart at $2\ \mathrm{ms^{-1}}$. After the collision Isobel is stationary. Find
(i) Stuart's velocity after the collision;
(ii) the coefficient of restitution between the cars;
(iii) the impulse acting on Stuart's car;
(iv) the kinetic energy lost in the collision.

6. A trapeze artist of mass $50\ \mathrm{kg}$ falls from a height of $20\ \mathrm{m}$ into a safety net.
(i) Find the speed with which she hits the net. (You may ignore air resistance and should take the value of g to be $10\ \mathrm{ms^{-2}}$).

Her speed on leaving the net is $15\ \text{ms}^{-1}$.

(ii) What is the coefficient of restitution between her and the net?

(iii) What impulse does the trapeze artist receive?

(iv) How much mechanical energy is absorbed in the impact?

(v) If you were a trapeze artist would you prefer a safety net with a high coefficient of restitution or a low one?

7. In each of the situations (a) – (f), a collision is about to occur. Masses are given in kilograms, speeds are in ms^{-1}.

(i) Draw diagrams showing the situations before and after the impact, indicating the values of any velocities which you know, or the symbols you are using for those which you do not know.

(ii) Write down the equations corresponding to the law of conservation of momentum and to Newton's Law of Impacts.

(iii) Find the final velocities.

(iv) Find the loss of kinetic energy during the collision.

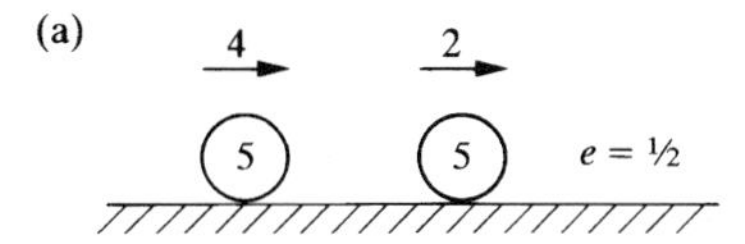

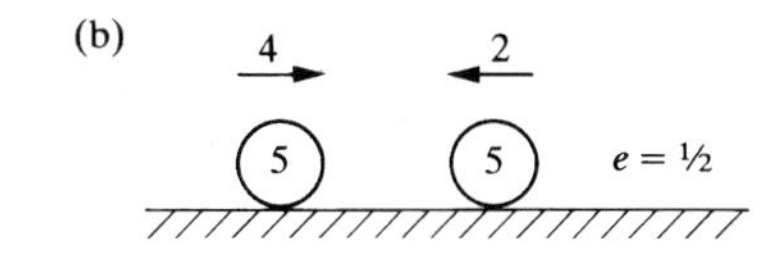

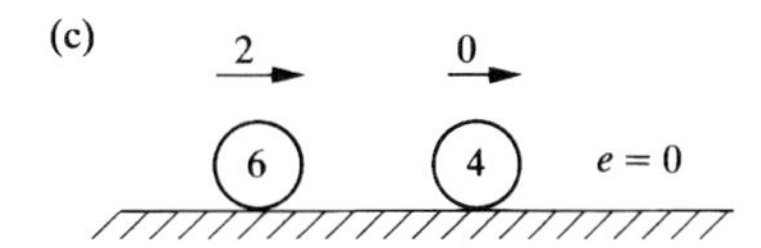

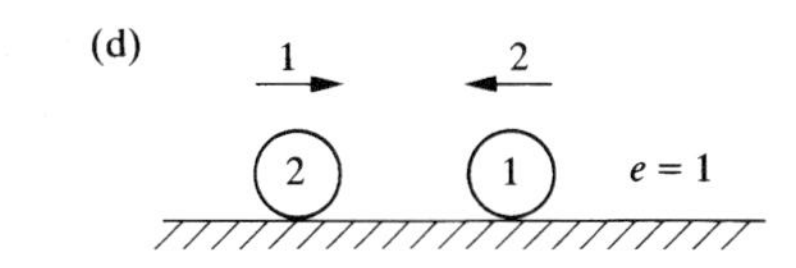

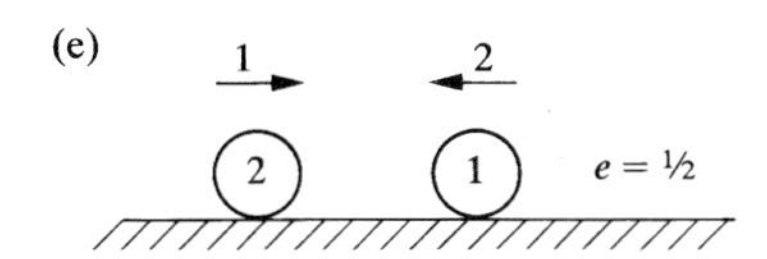

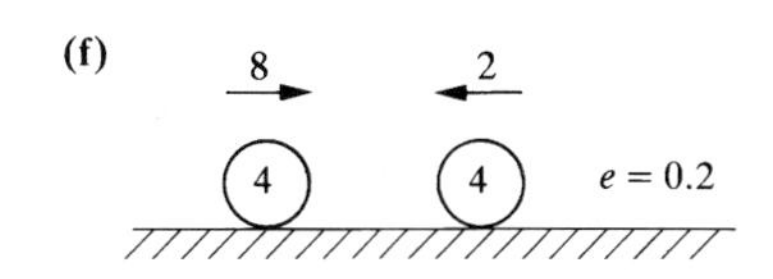

8. Three identical spheres are lying in the same straight line. The coefficient of restitution between any pair of spheres is $\frac{1}{2}$. The left hand ball is given speed $2\ \text{ms}^{-1}$ towards the other two. What are the final velocities of all three, when no more collisions can occur?

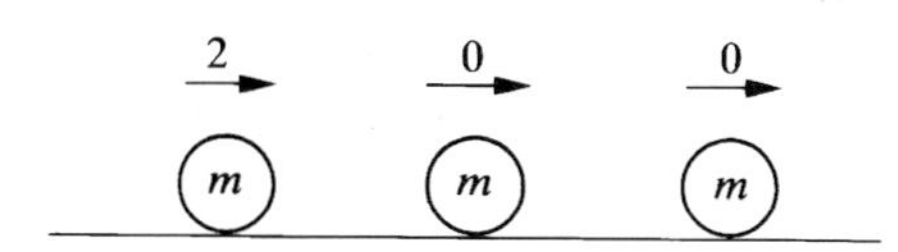

Exercise 5C continued

9. The diagram shows two snooker balls and one edge cushion. The coefficient of restitution between the balls and the cushion is 0.5 and that between the balls is 0.75. Ball A (the cue ball) is hit directly towards ball B with speed 8 ms^{-1}. Find the speeds and directions of the two balls after their second impact with each other.

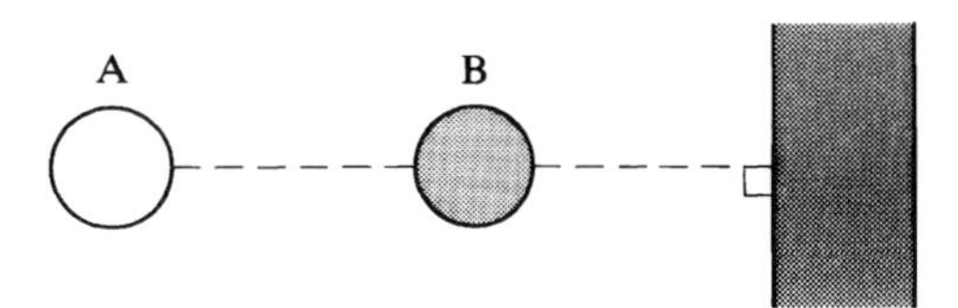

10. A sphere, A, of mass 40 g travels with speed 3 ms^{-1} along the line of centres towards another sphere, B, of mass 60 g which is initially stationary. After the collision sphere A is stationary.

(i) Draw diagrams to illustrate the situation before and after the collision.

(ii) Find the velocity of sphere B after the collision.

(iii) Find the coefficient of restitution between the spheres.

This situation may be generalised so that sphere A, of mass m_1, moving with speed u_1 along the ling of centres collides with the stationary sphere B, of mass m_2. After the collision sphere A is stationary. The coefficient of restitution between the spheres is e.

(iv) Show that $e = m_1 / m_2$.

(v) What does this tell you about

(a) the relative mass of the two spheres;

(b) the initial speed of sphere A?

Find the ratio of the kinetic energy after impact to that before impact and write it in terms of e. What must be the value of e if kinetic energy is to be conserved when the spheres collide?

11. The coefficient of restitution between a ball and the floor is e. The ball is dropped from a height h. Air resistance may be neglected, and your answers should be given in terms of e, h, g and n, the number of bounces.

(i) Find the time it takes the ball to reach the ground and its speed when it arrives there.

(ii) Find the ball's height at the top of its first bounce.

(iii) Find the height of the ball at the top of its nth bounce.

(iv) Find the time that has elapsed when the ball hits the ground for the second time, and for the nth time.

(v) Show that according to this model the ball comes to rest within a finite time having completed an infinite number of bounces.

(vi) What distance does the ball travel before coming to rest?

(vii) At what point do you think the model breaks down?

Exercise 5C continued

12. Two spheres of equal mass, m, are travelling towards each other along the same straight line when they collide. Both have speed v just before the collision and the coefficient of restitution between them is e. Your answers should be given in terms of m, v and e.

(i) Draw diagrams to show the situation before and after the collision.
(ii) Find the velocities of the spheres after the collision.
(iii) Show that the kinetic energy lost in the collision is given by $mv^2(1-e^2)$.
(iv) Use the result in part (iii) to show that e cannot have a value greater than 1.

Investigations

Superballs

What happens if a small superball is placed on top of a larger superball, as shown in the diagram, and both balls are dropped together.

A bouncing ball

Roll a ball off a table and mark the places on the floor where it bounces. Do these give a consistent value for the coefficient of restitution between the ball and the floor?

You may find it helpful to roll the ball down a fixed slope before it leaves the table, as shown in the diagram, and to sprinkle talcum powder on the floor so you can see where it lands, or to wet it and have it bounce on a sheet of sugar paper.

KEY POINTS

- The impulse from a force **F** is given by **F***t* where ***t*** is the time for which the force acts. Impulse is a vector quantity.
- The momentum of a body of mass m travelling with velocity **v** is given by m**v**. Momentum is a vector quantity.
- The impulse-momentum equation is
  Impulse = change in momentum.
- The law of conservation of momentum states that when no external forces are acting on a system, the total momentum of the system is constant. Since momentum is a vector quantity this applies to the magnitude of the momentum in any direction.
- Coefficient of restitution $e = \dfrac{\text{speed of separation}}{\text{speed of approach}}$

6 Frameworks

I have yet to see any problem, however complicated, which, when looked at in the right way, did not become still more complicated.

Paul Anderson

The structures in these photographs are similar: each is a framework made up of triangular elements. You will have seen many other structures like these. Why is the triangle the basic element in so many structures?

This chapter is concerned with the forces in structural frameworks. Not all structures are based on frameworks, but many are because the strength to weight ratio for a framework is usually higher than for a solid structure.

A framework is an arrangement of structural members (rods or cables). Frameworks are almost always made from triangular elements fitted together in two or three dimensions, because a triangle is rigid even if it is freely hinged (or *pin-jointed*) at the corners. If any other shape, such as a quadrilateral, were used, the joints would have to be rigid. This would make both the design of the structure and its successful construction far more difficult.

To analyse the forces in frameworks is quite complex, unless you can make two simplifying assumptions:

(i) that all members of the framework are light, so that their masses can be ignored;
(ii) that all the joints are smooth pin joints. This means that the bars could rotate about the joints without any resistance: these are no moments acting at the joints. All the forces are directed along the rods.

The consequence of these assumptions is that all the internal forces are directed along the rods.

These assumptions mean that the only forces that need to be considered are the external forces on the framework, and the tensions or thrusts in each rod of the framework.

For Discussion

ABCDE is a framework of seven light freely jointed rods, all of the same length, supported at A and D. A weight of 1000 N is hanging from E.

Which of the rods do you think are in tension and which in compression (thrust)?

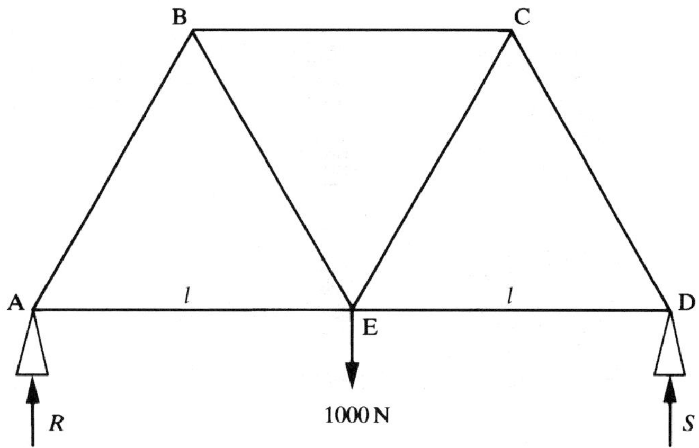

Now follow the analysis of this structure in the worked example below and see whether your intuitive ideas were correct.

EXAMPLE

Calculate the forces in each of the rods in the structure above, stating whether the rod is in tension or compression.

Solution

EXTERNAL FORCES

There are three external forces acting on the framework, R N and S N vertically upwards and 1000 N vertically downwards. Looking at the equilibrium of the structure as a whole:

For vertical equilibrium, $R + S = 1000$

Taking moments about A:

$\Rightarrow$ $$S = 500, \quad R = 500$$

(Notice that you could have found this result by symmetry).

INTERNAL FORCES

The next step is to mark the internal forces on all the rods. In this case they are all unknown and have all been marked as tensions (figure 6.1); if in fact they are thrusts they will be found to have negative values.

The tensions (in newtons) have been called $T_1, T_2, T_3, \ldots T_7$. They could equally have been called $T_{AB}, T_{BE}, \ldots T_{CD}$.

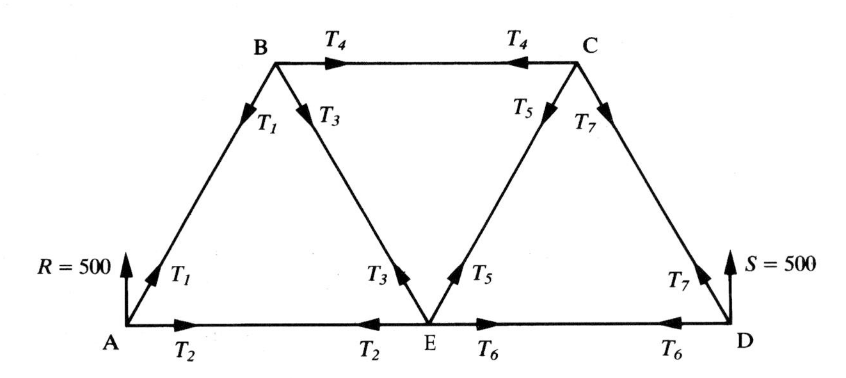

Figure 6.1

Now consider the equilibrium of each joint in turn. It is best to avoid joints where there are more than two unknowns so start at A or D.

Starting with A (figure 6.2):

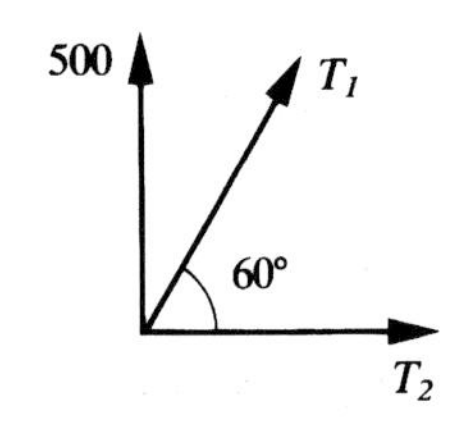

Figure 6.2

Vertical equilibrium: $T_1 \sin 60° + 500 = 0$

$\Rightarrow \quad T_1 = -577$ (thrust).

Horizontal equilibrium: $T_1 \cos 60° + T_2 = 0$

$-577 \cos 60° + T_2 = 0$

$\Rightarrow \quad T_2 = 289$ (tension).

Moving on to point B (figure 6.3):

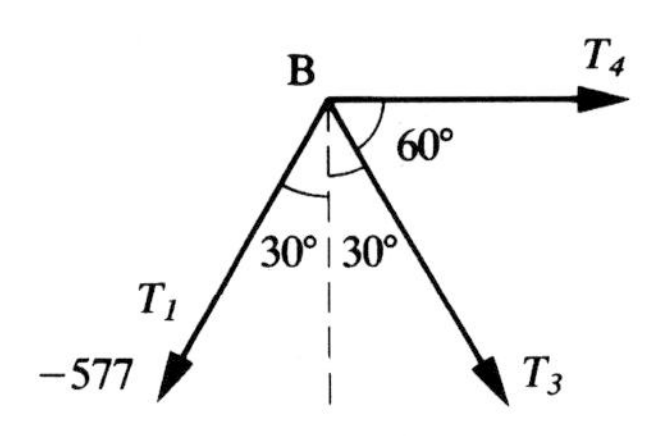

Figure 6.3

Vertical equilibrium: $T_1 \cos30° + T_3 \cos30° = 0$

$\Rightarrow$ $-577\cos30° + T_3 \cos30° = 0$

$\Rightarrow$ $T_3 = 577$ (tension).

Horizontal equilibrium: $T_4 + T_3 \cos60° - T_1 \cos60° = 0$

$\Rightarrow$ $T_4 + 577\cos60° + 577\cos60° = 0$

$\Rightarrow$ $T_4 = -577$ (thrust).

The three remaining tensions can be deduced because both the framework and the external forces are symmetrical about the vertical through E.

$T_7 = T_1 = -577$ (thrust) ; $T_5 = T_3 = 577$ (tension); $T_6 = T_2 = 289$ (tension).

Alternatively you can find them by considering the equilibrium of the other joints.

Figure 6.4 shows the forces in all of the members.

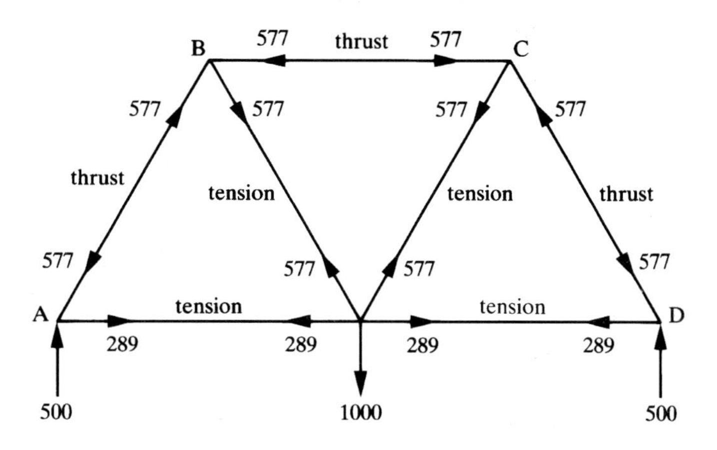

Figure 6.4

Experiment

The illustration shows a framework formed from an experimental structures kit.

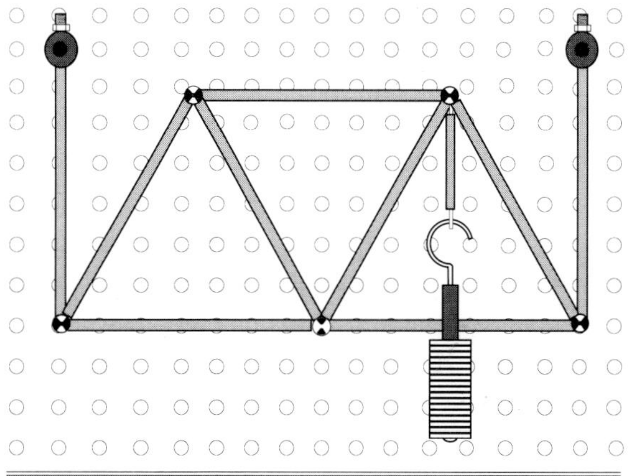

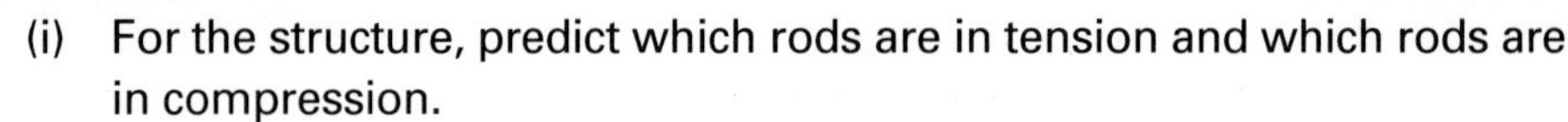

(i) For the structure, predict which rods are in tension and which rods are in compression.
(ii) Which rods could be replaced by strings?
(iii) You will see that there is a load applied to the structure. Calculate the force in each rod when the mass of the load is 200 g.
(iv) Make the structure from a kit, or other suitable equipment. Replace each rod in turn with a force meter to check your predictions.

Exercise 6A

1. The diagram shows a framework made up of three light rods which are freely jointed at A, B and C. A load of 2000 N is applied at the point B.

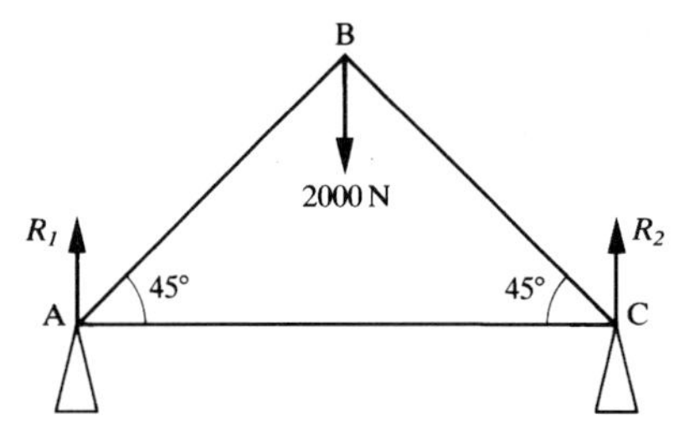

(i) Find the magnitudes of the forces R_1 and R_2.
(ii) Draw a sketch of the framework and mark in tensions T_1, T_2 and T_3 acting in the rods AB, BC and CA respectively.
(iii) If one of these rods is actually in compression, how will the value of the tension show this?
(iv) Write down equations for the horizontal and vertical equilibrium of joint A and solve these equations to find T_1 and T_3.
(v) By considering the equilibrium of joint B, find T_2.
(vi) Show that with the values of T_2 and T_3 which you have found, joint C is also in equilibrium.
(vii) Write down the forces in the three rods, stating whether they are in tension or in compression.

2. A framework LMN consists of three light, freely jointed rods LM, MN and NL. Their lengths, in metres, are as shown in the diagram. The framework is suspended from the points L and M by vertical strings with tensions S_1 and S_2 newtons and a weight of 1500 N is hung from N. The rod LM is horizontal.

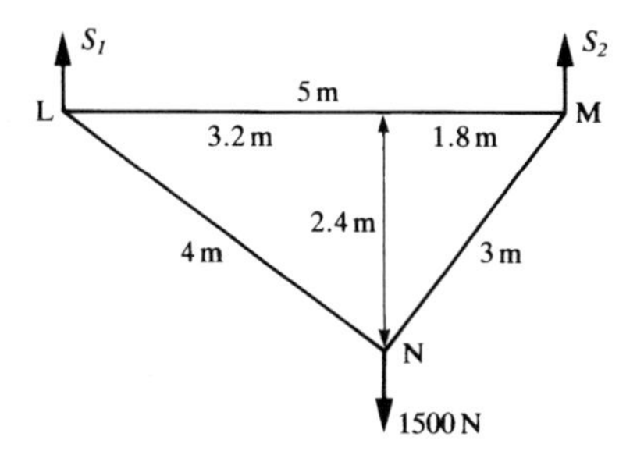

Exercise 6A continued

(i) Find the tensions S_1 and S_2.

(ii) Draw a sketch of the framework and mark in tensions T_1, T_2 and T_3 acting in the rods LM, MN and NL respectively.

(iii) Write down equations for the horizontal and vertical equilibrium of point M and solve these equations to find T_1 and T_2.

(iv) By considering the equilibrium of point N, find T_3.

(v) Show that with the values of T_1 and T_3 which you have found, point L is also in equilibrium.

(vi) Write down the forces in the three rods, stating whether they are in tension or in compression.

3. Two 100 N weights are suspended from the points B and C of the light pin-jointed framework ABC shown. The angles ABC and ACB are both 30°. The framework is itself suspended by a light string attached to A.

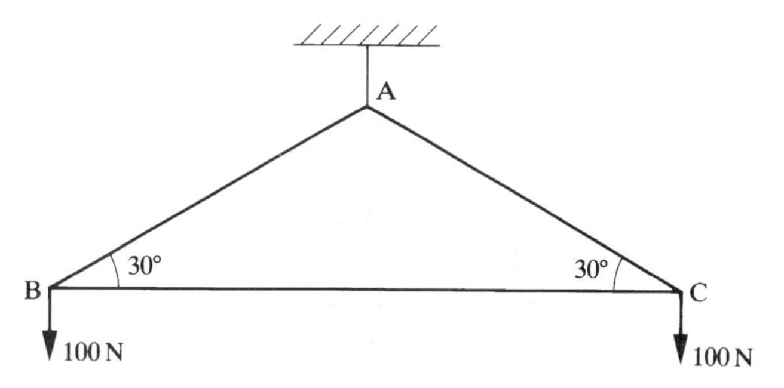

(i) Write down the three external forces acting on the framework.

(ii) Find the internal forces in the three rods, AB, AC and BC and state whether they are in tension or compression.

4. The diagram shows a framework PQR of light, freely jointed rods in the shape of an equilateral triangle. The framework is freely hinged to the wall at P. A light string connecting Q to S is taut when QS and PR are horizontal, as in the diagram. A weight of 500 N is hanging from R.

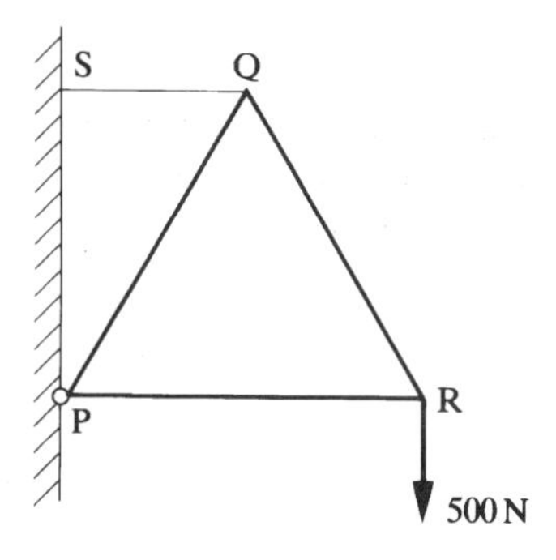

(i) By considering the equilibrium of the whole framework PQR, find the tension in the string QS.

Exercise 6A continued

(ii) Find the reaction of the wall on the framework at P
 (a) in the horizontal and vertical components,
 (b) in magnitude and direction form.

(iii) Find the internal forces in the three rods, stating whether they are in tension or in compression.

5. The diagram shows a simple crane. AB is a cable; BC, CD and DB are light freely jointed rods. The framework is freely hinged to its base at D. A load of 5000 N is hanging from C. The rod CD and the cable BA both make angles of 45° with the horizontal and BC is horizontal.

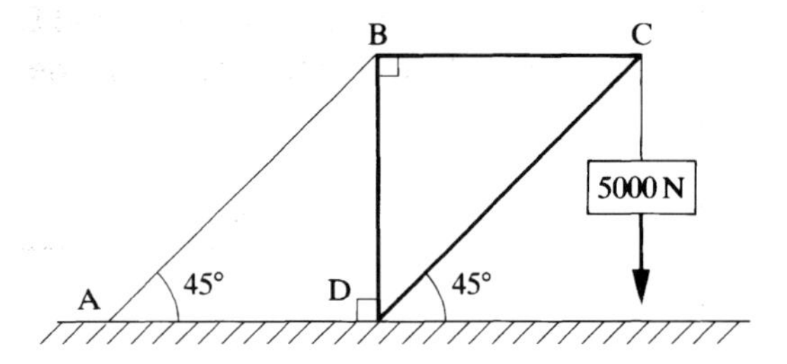

(i) Find the tension in the cable AB.

(ii) Find the magnitude of the reaction of the base on the framework at D.

(iii) Find the force in each rod, stating whether it is in tension or compression.

6. The diagram shows a crane supporting a load of 2000 N. The framework is light and freely jointed, and is secured at points A and C.

(i) Find the external forces acting on the framework.

(ii) Find the force in each member of the framework stating whether it is in tension or compression.

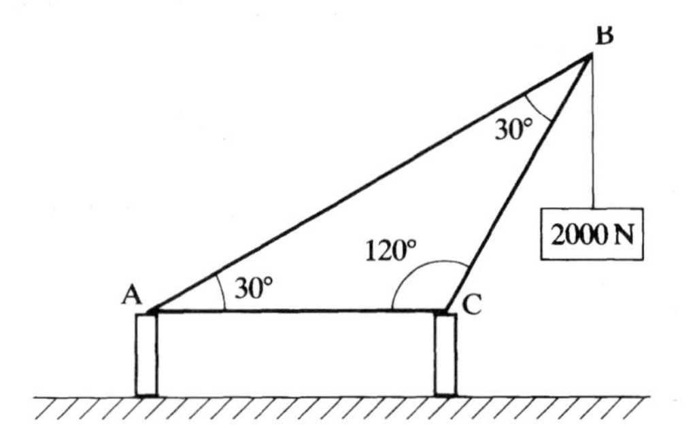

7. The diagram shows a framework of light, freely jointed rods supported at its base and carrying a load of 2500 N at its top. Find the force in each of the five rods, stating whether it is in tension or in compression. You may have seen a structure like this in use. Where?

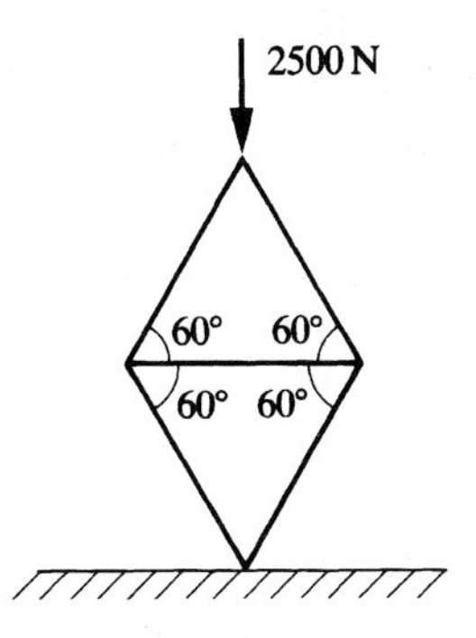

8. The light freely jointed framework shown below is supported at E and C and is carrying loads at A, B and D. Find the magnitudes of the forces in each of the seven rods and say whether they are in tension or compression.

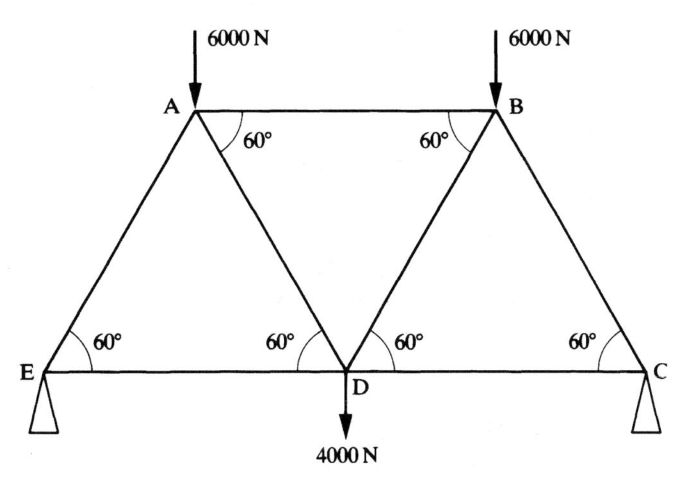

9. The light, freely jointed framework shown below is supported at E and C and is carrying loads at A, B and D. Find the magnitudes of the forces in each of the seven rods and state whether they are in tension or in compression.

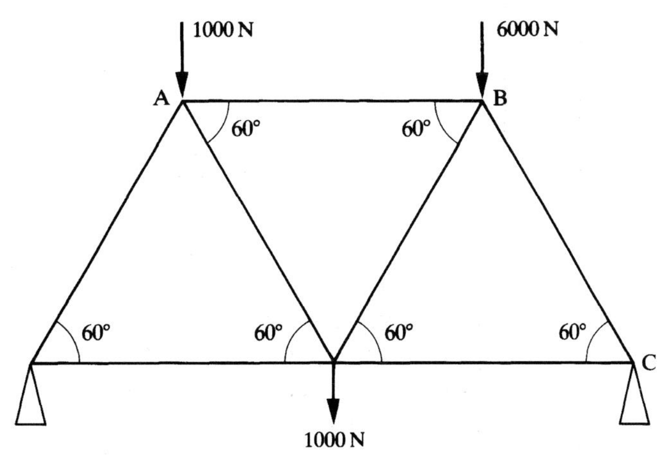

Exercise 6A continued

10. The light, freely jointed framework shown below is supported at S and T and is carrying loads of 400 N at both P and R as shown. Find the magnitudes of the forces in each of the seven rods and state whether they are in tension or compression.

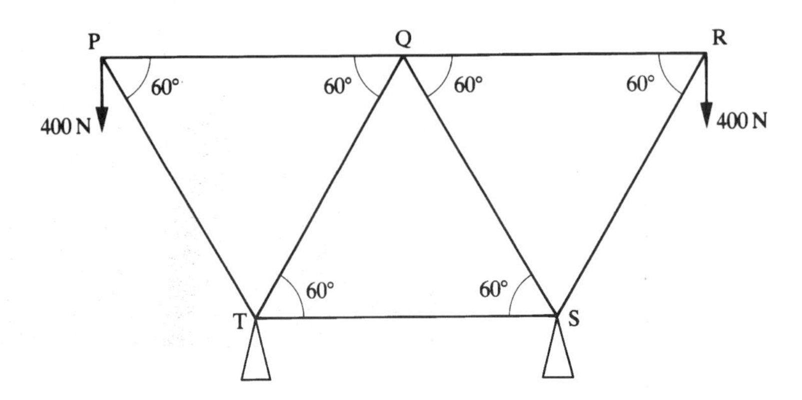

Investigations

Make two and three dimensional structures from a kit and investigate the forces acting in your structures.

The Forth Railway Bridge

The photograph shows a living model of the Forth Bridge at the time of its construction: "the chairs a third of a mile apart, the mens' heads 360ft above ground". Investigate this structure.

A 3-dimensional framework

The photograph shows a common type of clothes drier. Investigate its design and estimate the forces in the rods and strings when it is loaded in various ways.

KEY POINTS

- To calculate the forces in the members of a framework, the equilibrium of each joint must be considered.
- Calculations are usually simplified by making these two assumptions:
  1. all the members of the framework are light rods;
  2. all the joints are smooth and so there are no moments acting at them.
- First calculate the external forces acting on a framework:
  1. by considering horizontal and vertical equilibria;
  2. by taking moments about a suitable point.
- Then calculate the internal forces making these two assumptions:
  1. all the members of the framework are light rods;
  2. all the joints are smooth pin - joints.

  Consider the horizontal and vertical equilibrium.

Answers to selected exercises

1 A MODEL FOR FRICTION

Exercise 1A

1. (i) 0.1 (ii) 0.5
2. (i) does not move (ii) 3.675 ms^{-2} (iii) 2.205 ms^{-2} (iv) does not move
3. 4.802 kN
4. (i) 1.02 ms^{-2} (ii) 0.102 N (iii) 0.104 (iv) mass
5. 0.408
6. (i) smoother contact (ii) 0.2 (iii) 137.2 N
7. (i) 7.35 ms^{-2} (ii) 31.8 ms^{-1} (iii) 104.5 m
8. (i) 58.8 N (ii) 67.9 N
9. (i) 0.7 (ii) 35° (iii) 2.14 (iv) 50.2°
10. (ii) 4.42 ms^{-2} (iii) 5.15 ms^{-1} (iv) 5.41 m
11. greater than, equal to, less than 16.7° respectively
12. (ii) 73.3° (iii) 24.95 N
13. (i) 5.88 N (ii) it rests in equilibrium
14. (i) 0.194 (ii) 4.84 ms^{-2} (iii) 9.84 ms^{-1} (iv) 12.1 ms^{-1}
15. (i) (a) 37.9 N (b) 37.2 N (c) 37.5 N (ii) $40/(\cos\alpha + 0.4\sin\alpha)$ (iii) 21.8°

2 MOMENTS OF FORCES

Exercise 2A

1. (a) 15 Nm (b) −22 Nm (c) 18 Nm (d) −28 Nm
2. (a) 2.1 Nm (b) 6.16 Nm (c) 0.1 Nm (d) 0.73 Nm
3. 28.6 N, 20.4 N
4. 96.5 N, 138.5 N
5. (i) 1225, 1225 N (ii) 1449 N, 1785 N
6. 825 N, 775 N
7. (i) P = 27.5 g N, d = 147.5 g N (ii) P = 2.5 g N, Q = 172.5 g N (iii) If child moves to within 0.95 m of the adult the bench will tip (iv) the bench tips
8. (i) 15 g N, 30 g N (ii) 90 g N, 5 g N
9. (i) 0.5 g (30 − x) kN, 0.5 g (20 + x) kN
10. (i) 35 g N, 75 g N (ii) no (iii) 15.8 kg
11. (i) 2262 N, 7538 N (ii) 6 (iii) 784 N
12. (i) 3600 N (ii) 0.0017 m^{-2} (iii) 101 N

Exercise 2B

1. (i) 6 Nm (ii) −10.7 Nm (iii) 23 Nm (iv) 0 (v) −4.24 Nm (vi) 4.24 Nm
3. David and Hannah
4. (i) 5915 N (ii) 9062 N (iii) 4698 sec θ N
5. (i) 42.4 N (ii) 27.7 N (iii) 30.05 N
6. (iii) 30 Nm (v) 8.04 N, 15.40 N (vi) (a) 33.7° (b) 3.23 m
7. (i) 1405 N (ii) 638 N, 1611 N
9. (ii) 0 (iii) 141 N (iv) 141 N (v) $\mu \le 0.289$ (vi)
10. (a) (ii) 56.6, 56.6, 19.6 (iii) 0.29 (b) (ii) 98, 98, 196 (iii) 0.5 (c) (ii) 26.3, 26.3, 19.6 (iii) 0.13
11. (ii) 16.2 N (iii) 617 N
12. (ii) 1 812 258 Nm
13. (ii) 0.25, 0.75 (iv) 3.6°, 2193, 2847 N (v) 10.6°
14. (i) (A) 19.4° (B) 36° (C) 49.1° (ii) $\phi = \arctan(2\tan\theta)$

Exercise 2C

1. (i) 27.44 N (ii) 34.3 N (iii) slide
2. (i) P = 19.6 μ (ii) 5.88 N (iii) μ <.3 (iv) μ >.3
3. It slides
4. (i) 480 N (ii) 600 N (iii) slides
5. (a) (i) 450 (ii) 439 (iii) Topples (b) (i) 455 (ii) 518 (iii) slides 63°
6. (i) 14.0° (ii) 18.4° (iii) sliding
7. (i) (A) 50 x 20 (B) 10 x 20 (iii) (A) μ <.2 (B) μ > 5

3 CENTRE OF MASS

Exercise 3A

1. (i) 0.2 m (ii) −0.72 m (iii) + 0.275 m (iv) 1.19 m (v) 0 (vi) −0.92 m (vii) 36 cm (viii) 0.47 m
2. 2.18 m from 20 kg child
3. 4.2 cm
4. $1\frac{1}{3}$ m from hook
5. 0.83 m from base
6. 11.9 cm
7. $3\frac{1}{3}$ mm from centre
8. 4680 km
9. 3 cm
10. 2.4375 m from left
11. $m_2 l/m_1 + m_2)$ from m_1 end
12. (i) −1.85 (iv) $\dfrac{2Mad}{(l-d)(a-d)}$

Exercise 3B

1. (i) $(2\frac{1}{2}, \frac{5}{6})$ (ii) (0,2) (iii) $(\frac{1}{24}, \frac{1}{6})$ (iv) (−2.7, −1.5)
2. $(5, 6\frac{1}{3})$
3. $(12, 17\frac{1}{4})$
4. (i) (20, 60) (ii) (30, 87.5) (iii) (30, 60)
5. 23 cm
6. From left (i) (1.5, − 1.5) (ii) (1.5, −2.05) (iii) (1.68, -2.5)
7. (i) (a) (10, 2.5) (b) (12.5, 5) (c) (15, 7.5) (d) (17.5, 10) (e) (20, 12.5) (ii) 5 (iii) 11 (iv) 205 cm
8. (i) (28,60) (ii) (52, 60) (iv) 40 cm
9. (0.5a, 1.2a), 3.9°, 2 m
10. (i) 2.25 (ii) 0.56 (iii) 0.40, $(\frac{1}{2}, 1\frac{1}{2})$
11. (i) (0,96.15) (ii) 24, 28.6
12. (i) (a) $\left(\dfrac{4m+M}{8m+4M}\right)h$ (b) $\dfrac{h}{2}\left(\dfrac{M\alpha^2+m}{M\alpha+m}\right)$

4 ENERGY, WORK AND POWER

Exercise 4A

1. (i) 2500 J (ii) 40 000 J (iii) 5.6×10^9 J (iv) 3.7×10^{28} J (v) 10^{-25} J
2. (i) 1000 J (ii) 1070 J (iii) 930 J (iv) None
3. (i) 4320 J (ii) 4320 J (iii) 144 N
4. (i) 540 000 J (ii) 3600 N

5. (i) 500 000 J (ii) 6667 N
6. (i) 64 J (iii) 64 J (iv) 400 N
7. (i) 3.146×10^5 J (ii) 8.28×10^{-3} N
8. 18.6 ms^{-1}
9. (i) 240 N (ii) 5.5 m (iii) 1320 J
10. (i) 4.472 ms^{-1} (ii) 4.5 J (iii) 4.243 ms^{-1} (iv) 0.25 J (v) 89.4 ms^{-1}

Exercise 4B

1. (i) 9.8 J (ii) 94.5 J (iii) −58.8 J (iv) −58.9 J
2. (i) −27.44 J (ii) 54.88 J (iii) −11.76 J
3. 17.64 J
4. 23 285 J
5. (i) 1 573 743 J (ii) 20 000 J
6. (i) (a) 1504 J (b) 280 J (ii) 15.6 ms^{-1}
7. (i) 2122 J (ii) b
8. (i) (a) 1701 J (b) 8.7 ms^{-1} (ii) (a) unaltered (b) decreased
9. (i) 2450 J (ii) 2450 J, 9.9 ms^{-1} (iii) 60°
10. (i) 153.7 J (ii) 153.7 − 1.96x (iii) $0 \le t \le 4$ (iv) 27.7 ms^{-1} (v) 28.8 m
11. (i) 34300 J, 21875 J (ii) 248.5 N (iii) 5061 N
12. (i) 9.8 ms^{-2} (ii) $1.47\,(10t - 4.9t^2)$ (iii) 5.1 m, 10**i** (iv) 10**i** + 10**j**, 14.1 ms^{-1}, 15 J
13. (i) 109 ms^{-1} (ii) 114 N
14. (i) 12 J, 9.408 J (ii) 2.592 J (iii) 0.162 N (iv) 15.07 ms^{-1}
15. (i) 5 m (ii) 6.26 ms^{-1} (iii) 8.94^+ m (9^+ m) from middle
16. (i) 591.6 J (ii) 758,700 J (iii) 210.75 W

Exercise 4C

1. (i) 308.7 J (ii) 37 044 J (iii) 10.29 J
2. (i) 2352 J (ii) 1176 W (iii) 1882 W, 0 W, 2822 W
3. (i) 31 752 J (ii) 16 200 J (iii) 1598 J (iv) 1332 N (v) Power
4. 703 N
5. (i) 611 N (ii) Mass of cyclist
6. 245 000 W
7. (i) 560 W (ii) 168 000 J
8. (i) 1253 J (ii) 209 W
9. (i) 20 ms^{-1} (ii) 0.0125 ms^{-2} (iii) 25 ms^{-1}
10. (i) 1.6×10^7 W (ii) 0.0025 ms^{-2} (iii) 5.7 ms^{-1}
11. (i) 16 (ii) 320 N (iii) 6400 N
12. (i) 19.27 (ii) 34.1 ms^{-1} (iii) 57.4 ms^{-1}

5 IMPULSE AND MOMENTUM

Exercise 5A

1. (i) 500 Ns (ii) 2000 Ns (iii) 2.8×10^8 Ns (iv) 2×10^{-22} Ns
2. (i) 18 200 Ns (ii) 0.00225 Ns (iii) 3 Ns (iv) 10 Ns
3. (i) 15 (ii) 20 (iii) 20 (iv) 300 J (v) 30 Ns
4. (i) 1.2 Ns (ii) 2 s (iii) 0 (iv) 1.2 Ns (v) 0.6 N
5. (i) 2.125 Ns (ii) (a) 21.25 N (b) 42.5 N
6. (i) 11 880 Ns (ii) 99 000 N (iii) 11 g
7. (i) 5.42 ms^{-1} (ii) 108.4 Ns (iii) 108.4 Ns (iv) 542 N
8. (i) + 30000 Ns (ii) − 15 000 Ns (iii) − 45 000 Ns (v) 4 500 000 N
9. (v) 0.32 Ns
10. (i) − 20**i** (ii) 5.1**i** + 0.60**j** + 0.450**k** (iii) Fatima
11. (i) 0.048 Ns (ii) 0.432 N
12. (ii) 12 250 Ns (iii) 4.9 ms^{-1}

Exercise 5B

1. 2 ms^{-1}
2. (i) 25.6 ms^{-1} (ii) 4444 Ns (iii) 4444 Ns
3. 195.45 ms^{-1}
4. (i) $1\frac{2}{3}$ ms^{-1} (ii) 1667 Ns
5. (i) 0.623 ms^{-1} (ii) 9.97 Ns (iii) 19 938 N
6. (i) 4990 ms^{-1} (ii) 5000.001 ms^{-1}
7. (i) 3 ms^{-1}
8. (i) 0 (ii) M:-0.75 ms^{-1}, A:1.05 ms^{-1} (iii) M:−52.5 Ns, A: + 52.5 Ns (iv) 0
9. (i) 1 ms^{-1} (ii) 9 ms^{-1} (iii) 21 ms^{-1} (iv) 33 ms^{-1}
10. (i) 10 ms^{-1} (ii) 24 000 Ns (iii) 24 000 Ns (iv) 960 000 N (v) 0.5 m
11. (i) $\frac{2}{3}$ ms^{-1}
12. (i) 49 000 J (ii) 7 ms^{-1} (iii) 4 ms^{-1} (iv) 42 A%

Exercise 5C

1. (i) $\frac{2}{3}$ (ii) 1.44 ms^{-1} (iii) $\frac{3}{4}$ (iv) 3.2 ms^{-1}
2. (i) 0.3 (ii) 0 (iii) 0.95 (iv) 1
3. (i) 0.8 (ii) 1.62 Ns (iii) 2.43 J
4. (i) 4.427 ms^{-1} (ii) 3.98 ms^{-1} (iii) 0.9 (iv) 0.149 J (v) 0.149 J (vi) 0.6561 m
5. (i) 1 ms^{-1} (ii) 0.2 (iii) 450 Ns (iv) 900 J
6. (i) 20 ms^{-1} (ii) $\frac{3}{4}$ (iii) 1750 Ns (iv) 4375 J
7. (a) (iii) $2\frac{1}{2}$, $3\frac{1}{2}$ ms^{-1} (iv) 3.75 J
   (b) (iii) $-\frac{1}{2}$, $2\frac{1}{2}$ (iv) 33.75 J
   (c) (iii) 1.2, 1.2 ms^{-1} (iv) 4.8 J
   (d) (iii) −1, 2 ms^{-1} (iv) 0 J
   (e) (iii) $-\frac{1}{2}$, 1 ms^{-1} (iv) 2.25 J
   (f) (iii) 2, 4 ms^{-1} (iv) 96 J
8. $\frac{13}{32}$, $\frac{15}{32}$, $\frac{9}{8}$
9. $\frac{47}{16}$, $\frac{7}{16}$
10. (ii) 2 ms^{-1} (iii) $\frac{2}{3}$
11. (i) $\sqrt{\frac{2n}{g}}$ $\sqrt{2gh}$ (ii) e^2h (iii) $(e^2)^{\mu}$ h
    (iv) $\sqrt{\frac{2h}{g}}\,(1 + 2e)$, $\sqrt{\frac{2h}{g}}\,(1 + 2e + \ldots 2e^{n-1})$
    (vi) $h\left(\frac{1+e^2}{1-e^2}\right)$
12. (ii) -eu, eu

6 FRAMEWORKS

Exercise 6A

1. (i) 1000 N, 1000 N (v) 1414 (c) (vii) 1414 (c), 1000(T)
2. (i) 540 N, 960 N (vi) 720 (c), 1300(T), 900(T)
3. (i) 100 N, 100 N, 200 N (ii) 200 N(T), 200 N(T), 173 N (c)
4. (i) 577 N (ii) 577, 500 N (iii) 764, 40.9° (to horizontal) (iv) 577(T), 577(T) 288.7 (c)
5. (i) 7071 N (ii) 11 180 N (iii) 5000(T), 7071(c), 5000(c)
6. (i) -1000 N, 3000 N (ii) 2000 N(T), 3464 N (c), 1732 N (c)
7. 1443 (c), 1443 (c), 1443 (T)
8. AE,BC 9238 (c); AD,BD 2309 (T); ED,CD 4619 (T); AB 5774 (c)
9. AE 3175 (c); AB 2598(c); BC 6062 (c); CD 3031 (T); DE 1588 (T); AD 2021 (T); BD 866 (c)
10. PT, SR 462(c); PQ,QR,TS 231 (c); TQ, QS 0